NATURAL ORDER

Essays on Law, Policy and Psychiatry
Peter Fritz Walter

Codependence
Coping with Addiction, Sadism and Abuse

Eight Dynamic Patterns of Living
Base Elements of True Civilization

Emotional Flow
A Holistic Approach to Healing Sadism

Love or Laws?
When Law Punishes Life

Minotaur Unveiled
A Historical Assessment of Adult-Child Sexual Interaction

Natural Order
Thesis, Antithesis and Synthesis in Human Evolution

Pedophilia Revisited
The Making of a Crime for Justifying Lacking Social Policy

The Commercial Exploitation of Abuse
A Study on Social Policy

The Legal Split in Child Protection
Overcoming the Double Standard

The Roots of Violence
Why Humans Are Not by Nature Violent

NATURAL ORDER

Thesis, Antithesis and Synthesis in Human evolution

Peter Fritz Walter

Published by Sirius-C Media Galaxy LLC

Business Filings Incorporated

108 West 13th St., Wilmington, DE 19801, USA

©2018 Peter Fritz Walter. Some rights reserved.

Essays on Law, Policy and Psychiatry, Vol. 6

Creative Commons Attribution 4.0 International License

This publication may be distributed, used for an adaptation or for derivative works, also for commercial purposes, as long as the rights of the author are attributed. The attribution must be given to the best of the user's ability with the information available. Third party licenses or copyright of quoted resources are untouched by this license and remain under their own license.

The moral right of the author has been asserted

Set in Avenir Light and Trajan Pro

Designed by Peter Fritz Walter

ISBN 978-1-983990-43-4

Publishing Categories
Science / Life Sciences / Evolution

Publisher Contact Information
publisher@sirius-c-publishing.com
http://sirius-c-publishing.com

Author Contact Information
pfw@peterfritzwalter.com

About Dr. Peter Fritz Walter
http://peterfritzwalter.com

About the Author

Parallel to an international law career in Germany, Switzerland and the United States, Dr. Peter Fritz Walter (Pierre) focused upon fine art, cookery, astrology, musical performance, social sciences and humanities.

He started writing essays as an adolescent and received a high school award for creative writing and editorial work for the school magazine.

After finalizing his law diplomas, he graduated with an LL.M. in European Integration at Saarland University, Germany, in 1982, and with a Doctor of Law title from University of Geneva, Switzerland, in 1987.

He then took courses in psychology at the University of Geneva and interviewed a number of psychotherapists in Lausanne and Geneva, Switzerland. His interest was intensified through a hypnotherapy with an Ericksonian American hypnotherapist in Lausanne. This led him to the recovery and healing of his inner child.

After a second career as a corporate trainer and personal coach, Pierre retired in 2004 as a full-time writer, philosopher and consultant.

His nonfiction books emphasize a systemic, holistic, cross-cultural and interdisciplinary perspective, while his fiction works and short stories focus upon education, philosophy, perennial wisdom, and the poetic formulation of an integrative worldview.

Pierre is a German-French bilingual native speaker and writes English as his 4th language after German, Latin and French. He also reads source literature for his research works in Spanish, Italian, Portuguese, and Dutch. In addition, Pierre has notions of Thai, Khmer, Chinese, Japanese, and Vietnamese.

All of Pierre's books are hand-crafted and self-published, designed by the author. Pierre publishes via his Delaware company, Sirius-C Media Galaxy LLC, and under the imprints of IPUBLICA and SCM (Sirius-C Media).

Evolution is an ongoing and open adventure that continually creates its own purpose in a process whose detailed outcome is inherently unpredictable. Nevertheless, the general pattern of evolution can be recognized and is quite comprehensible. Its characteristics include the progressive increase of complexity, coordination, and interdependence; the integration of individuals into multileveled systems; and the continual refinement of certain functions and patterns of behavior.

—FRITJOF CAPRA, THE TURNING POINT (1987), P. 313

The author's profits from this book are being donated to charity.

Contents

Introduction — 13
The Shortest Summary of Human History

Chapter One — 21
The Early Natural Order

Minoan Civilization — 21
The Egalitarian Society — 25
The Nonviolent Trobriands — 27
Yin-Yang Balance — 32
Pleasure, the Prime Regulator — 34
Early Love for Pleasure — 34
Lovers and Fuckers — 36
The Perversion of Pleasure into Violence under Patriarchy — 38
Pleasure and Intelligence — 41
Pleasure and Touch — 44
Pleasure and Violence — 49

Child Sexuality and its Detractors — 52
Children and Sex — 52
A Note on Child Trauma — 55
Adolescents and Sex — 56
No Sex Education Needed — 60
Child Sexual Abuse — 62

Pedoemotions and Pedoerotics — 63
Definition — 63
What is the Nature of our Emotions? — 64
What are Pedoemotions? — 66
What is Pedoerotics? — 67
What is Adult-Child Sex Like? — 68

References	72

CHAPTER TWO — 73
The Destruction of the Natural Order

CARTESIAN REDUCTIONISM	73
THE UPSURGE OF MORALISM	83
What is Moralism?	85
Moralism in Education	86
REPRESSION AND PERVERSION	87
Repression and Denial	87
Repression and Regression	88
Repression and Retrogradation	89
Repression and Projection	91
Cultural Perversion	91
Legislative Perversion	94
Religious Perversion	99
LOVE AND SPLIT-LOVE	102
THE DISINTEGRATION OF SEXUAL PARAPHILIAS	107
Gerontophilia	107
Pedophilia or Childlove	109
Boylove or Pederasty	113
On the Existence of Nepiophilia	116
Childlove in Literature	119
PARENT-CHILD CODEPENDENCE	124
The Popular Confusion	128
The Pitfall of Emotional Entanglement	132
EMOTIONAL ABUSE	138
Introduction	138
The Primary Abuse Etiology	141
THE OEDIPAL MOLD	144
What means Oedipus Complex?	144
Is the Oedipus Complex Universal?	146

CONTENTS

Criticism of the Theory	152
OEDIPAL CULTURE	157
Are Masturbating Children Better Citizens?	159
The Dogma of the Autoerotic Consumer Child	163
Intellect Boosting for Sexually Demanding Children	166
Qualifying Oedipal Castration as Child Abuse?	169
Rationality vs. Oedipal Mysticism	172
Oedipal Hero	175
MYSTICISM AND ATHEISM	178
Scientific Mysticism	178
Mystical Thinking vs. Functional Thinking	179
Mysticism vs. Spirituality	184
Mysticism, Insanity, Cruelty, Brutality, Perversion and Fascism	186
NARCISSISM	189
What is Narcissism?	190
How To Identify Narcissism?	193
Narcissism and Soul	195
The Origin of Narcissism	203
DENIAL OF COMPLEXITY	208
The Etiology of Fascism	208
Complexity and Simplicity	208
Complexity and Consciousness	210
Complexity and Child Abuse	213
The Denial of Erotic Complexity	215
The Denial of Children's Erotic Complexity	219
THE PLAGUE OF SADISM	221
The Etiology of Sadism	221
The Abuse Pattern	222
Sadism and Moralism	224
CONSPIRACY THINKING VS. CRITICAL THINKING	225
Generalities	226
Dangers of Conspiracy Thinking	227

The Biggest Secret	229
YOUTH FASCISM	239
First Example	240
Second Example	241
Third Example	242
Encrypted Fascism	

CHAPTER THREE — 249
The New Natural Order

THE EIGHT DYNAMIC PATTERNS OF LIVING	249
General	249
The Eight Patterns	251
THE HOLISTIC SCIENCE PARADIGM AND WORLDVIEW	261
A Matter of Terminological Correctness	261
Ancient Wisdom Traditions	264
Goethe's Color Theory	265
THE TWELVE BRANCHES OF THE TREE OF KNOWLEDGE	266
Science and Divination	267
Science and Energy	276
Science and Flow	280
Science and Gestalt	282
Science and Intent	285
Science and Intuition	288
Science and Knowledge	292
Science and Pattern	301
Science and Perception	308
Science and Philosophy	313
Science and Truth	318
Science and Vibration	327
THE TRUE RELIGIO	330
Generalities	330
The Inner Selves	331
Inner Child	332

CONTENTS

Inner Adult	333
Inner Parent	333
Inner Dialogue	334
Multidimensionality of the Psyche	335
Function of the Ego	335
TOWARD A SCIENCE OF LIFE	336
Emotional Flow	336
The Nature of Emotions	337
Emotional Awareness	340
Emotional Balance	341
Emotional Intelligence	343
The Life Force	344
The Emonics Terminology	346
PRIMARY POWER AND LIVE YOUR LOVE	353
Primary Power, Soul Power, Self Power	353
Live Your Love	356
PERMISSIVE EDUCATION	357
Introduction	358
The Failure of Moralistic Education	360
Raising Humane Humans	369

POSTFACE 371
Where Are We Now?

BIBLIOGRAPHY 375

PERSONAL NOTES 439

Introduction

The Shortest Summary of Human History

I have written the present book with an intention to bring order to my holistic research on violence as the ultimate perversion of the human setup. As my approach was topic-based, I was getting ahead through little chunks of insight that appeared to randomly elucidate seemingly unrelated domains of human behavior.

It was only after twenty-four years of virtually tapping in the dark that I eventually realized the basic interconnectedness of all my research topics. In the evening of my 54th birthday, listening to Olivier Messiaen's *La Nativité du Seigneur* I wrote, in a sudden flash of insight, the outline of the present book. It was really a mental flash. The outline was on paper after five minutes. And in only ten days, I finalized the entire book.

NATURAL ORDER

This was a surprising event as I had by no means intended to write a new book. My intention was to write a series of scholarly articles that would bring some order to my research glossary that was published since several years on the Internet, and that never got any attention from the public.

The holistic vision I received was not conscious, but underlying and subtle; it was a driving force for writing the book. The lucidity I experienced in that moment helped me to see clearly what was highest priority and mere secondary in my research, thus helping me to eventually focus on the main topics, instead of wasting my time with the little critter.

I believe that human written history is *truncated* in the sense that it has never presented a valid and complete outline of human behavior. It is, if ever, an example of how ideologies, religious or political, can forge a science of history, and all science, in a world where truth, if not politically correct, is felt as a threat to human togetherness.

We have learnt through quantum physics that not objects, elements, stuff and matter are of importance, *but how we look at them*. On the subatomic level, all is movement and potential, but I do believe this is even true in our everyday reality. What we give attention to, we increase and strengthen, what we overlook tends to disappear not only from our regard, but even physically, in some way or the other, from our lives. Hence the need to give attention to detail, which is love!

I believe that within our patriarchal societies, this attention to detail was given too much to things unrelated to the human body. It was given to deities, concepts and ideas, to religions and 'holy scriptures,' to grandiose promises of Heil and Salvation, to gurus and saviors, to warriors and crusades against this or that, but it was hardly ever given to what is *natural and plain*, to happiness, to the love play, to sensual bonding, to feelings, to friendship, and to good sex with the right partner at the right moment, and in the right mood …

The natural order, as long as it was kept by female door keepers, was a functional one; when the door keepers changed and became males, decay set in, and stupidity. Behold, I am a male, so you won't be able to dismiss me as a hate-bitch or feminist! I am a man and yet I know that man's wisdom is inferior to woman's, man's strength inferior to woman's, and man's endurance inferior to woman's. We owe woman that we are born, for who did carry us as infants … , our fathers?

And yet, when you read most scriptures from most religions, it's all about men and seldom about women, while all those men were all born and breastfed (or not) by women. It's as if they wanted to blind out that they once had mothers and loved them, and perhaps loved them too much, and felt ashamed for that early love?

This book is about *intimacy*, in a very large sense, intimacy as the antidote to patriarchy, intimacy with ourselves, first of all, when we enter that inner dialogue and listen to our shadow, or our inner child.

Intimacy was shunned by patriarchy, and today it's one of the things that are under tight supervision of our police departments, our secret intelligence services, our prosecutors, and our neighbors as the spy-frontline in an altogether paranoid theater of global dimensions!

This book tells a story, the story of intimacy, as a *thesis*, an *antithesis* and, hopefully, a *synthesis* to come. The thesis was matriarchy, the antithesis was patriarchy, and the synthesis has no name. It is potential reality, a *quantum field* at best.

I think this book is queer and uncomfortable, not easily digested, not to be read on full stomach. I almost wanted to add … and not for minors!

When you begin reading, you hopefully question what you know of intimacy between *majors and minors*, for example, or the reason why minors are called *minors*, in the first place.

Sounds like … of *minor* value, right?

Legally speaking, children are of minor value, really. Their consent to pleasure and sex is deemed 'legally invalid' in criminal law; and in correction homes for children no human rights declarations have been enacted so far, only for prisons and jails for adults.

You didn't know that? Then it's a good time to learn about it, and a few other things, among them that queer construct called the 'Oedipus Complex' that serves as a fantastic candy for surprise parties. Simply because everybody seems to know about it while everybody knows to know nothing about it. This is then what is called a *myth*.

NATURAL ORDER

And there are more myths around in our glorious postmodern international consumer world, for example the myth that sex causes trauma in the brains and hearts of children, or that infants, when beaten or circumcised right after birth, do not really feel anything because 'they have deserved that.'

Or the myth that every boy is deeply in love with his mother and every girl deeply in love with her father, and that this love is 'of course' only platonic.

In the *world of parties and of myths* that is our standard day-and-night soup, in a standard world, that has forgotten about nature, about women, about love, and about children's genitals, this must be so. Because what is below the belt is 'private,, right? So private obviously that it can be mutilated by circumcision, but not touched for giving pleasure, as the first is considered *necessary for hygiene* and the second 'a crime.' Does that make sense to you? Perhaps. Not to me.

This book can't do away with all accumulated myths in a deeply irrational society that is pervaded by scientific and psychoanalytic mysticism. What this book well can do is to point a finger and say 'this is myth' and 'this is reality,' and when you apply

THE SHORTEST SUMMARY OF HUMAN HISTORY

quantum physics, you will have an easy way out from feeling responsible for your mess, right? You will say, oh, this is just the 'observer perspective' of the author. 'He's simply a queer looker, he sees it all with his black glasses, that's why it all sounds so black.'

You may do that, think that and believe that. But in that case, and that is why I write it here, I suggest you to read no further and throw this book far away, as far as possible, so that it may not cause you discomfort in your holy sleep!

I have hope that all scientists around the world will recognize the cosmic information field, and eventually start to see life as what it is, *subtle energy*, not a box of old brownish papers that smell myrrh, and on which is written 'God created the world in 7 Days' and so on.

You know the story.

And by the way, if you think that this introduction is written in colloquial language and looks strangely unfit for the rest that follows, then I congratulate you to your habitual lucidity that reveals you *subtle truth*.

CHAPTER ONE

The Early Natural Order

Minoan Civilization

The ancient Minoan culture from Crete was a highly developed civilization with a natural focus on sensuality, on beauty, free sexuality and a matriarchal worldview.

This culture had integrated the *Eight Dynamic Patterns of Living* that I found to be *regulatory* when it comes to evaluating why a given culture lives in a peaceful and integrated or else, in a violent and disruptive way.

Minoan culture respected all of the patterns I found to be essential for peaceful living and it had fully integrated the female, and also the female child in a partnership paradigm of living, and shared responsibility.

NATURAL ORDER

> —More information can be found in Riane Eisler's books and the many references they contain on Minoan Culture, ancient matriarchies and the perennial Goddess cult. See, for example, Riane Eisler, The Chalice and the Blade (1995) and Sacred Pleasure (1996).

No slavery and no physical punishment for children in schools was practiced. The crime rate in that culture was very low. Their religion did not worship a male god but a plethora of goddesses and spirits of nature. The low level of violence in that culture was exemplary in history, yet this civilization was virtually raped and devoured by the cruel, slavery-practicing invader tribes.

Riane Eisler, in her concise exposé on Minoan mores, culture and lifestyle in her book *The Chalice and the Blade (1995)*, speaks of Crete as 'the essential difference' and reminds that already Plato described the Minoans as 'exceptionally peace-loving people.'

Among the positive aspects Eisler lists about Minoan culture, referencing many other scholars, the most striking is that this ancient culture had a well-built model of what today we call *democracy*.

> Especially fascinating is how our modern belief of government should be representative of the interests of the people seems to have been foreshadowed in

THE EARLY NATURAL ORDER

> Minoan Crete long before the so-called birth of democracy in classical Greek times. Moreover, the emerging modern conceptualization of power as responsibility rather than domination likewise seems to be a reemergence of earlier views.
>
> —Riane Eisler, The Chalice and the Blade (1995), p. 38.

Among the features of Minoan civilization is the fact that although it was ruled by a centralized government, this government was not exploiting or brutalizing the masses, which clearly is an exception when we look at other civilizations of that time. And Eisler equally observes that while there was an affluent ruling class in Crete, there is no indication that it was 'backed up by massive armed might.'

> A remarkable feature of Cretan culture is that there are here no statues or reliefs of those who sat on the thrones of Knossos or of any of the palaces. Besides the fresco of the Goddess - or perhaps a queen/priestess - at the center of a gift-bearing procession, there seem to be no royal portrayals of any kind until the latest phase. Even then, the sole possible exception, the painted relief sometimes identified as the young prince, shows a long-haired youth, unarmed, naked to the waist, crowned with peacock plumes and walking among flowers and butterflies. (Id., p. 37)

Still today, the health of the Cretan population and their wistful lifestyle is famed and acknowledged. In fact, a recent demographic survey has shown that in Europe, the Cretan population is by far the healthiest in that *cancer and heart disease rates* are among the lowest in Europe, and even in the world at large.

Among modern scholars, Terence McKenna and Riane Eisler stand out in their correct evaluation of the value of Minoan civilization and its example status for modern peace research. In *Food of the Gods (1992)*, McKenna writes:

> The ambiance of Minoan-Mycenaean religion was one of realism, a sense of the vitality of bios, and sensual celebration. The snake-handling Minoan nature Goddess is representative of all these values. In all Minoan depictions, her breasts are full and bare and she handles a golden snake. Scholars have followed shamanic convention and have seen in the snake a symbol of the soul of the deceased. We are dealing with a goddess who, like Persephone, rules over the underworld, a shamaness of great power whose mystery was already millennia old. (Id., p. 127)
>
> In the age of kingship, only Crete—an island and in those times remote from the events of Asia Minor—harbored the old partnership model.
>
> The mysterious Minoan civilization became the inheritor of the style and gnosis of forgotten and far-off times. It was a living monument to the partnership ideal, enduring for three millennia after the triumph of the dominator style was everywhere else complete. (Id.)

THE EARLY NATURAL ORDER

The Egalitarian Society

Early research on matriarchy was often criticized with the pseudo-logical argument that overcoming patriarchy and going back to matriarchy would equal to grant women a dominator status over men.

> —The Swiss anthropologist and sociologist Johann Jakob Bachofen (1815-1887) is credited with the theory of matriarchy, or Mutterrecht, title of his major publication, a book that presented a radically new regard on the role of women in a broad range of ancient societies. Bachofen demonstrated that motherhood is the source of human society, religion, morality, and decorum and he drew upon Crete, Greece, Egypt, India, Central Asia, North Africa, and Spain. See Johann Jakob Bachofen, Gesammelte Werke, Band II: Das Mutterrecht (1948).

Riane Eisler, in her books *The Chalice and the Blade (1995)* and *Sacred Pleasure (1996)*, suggests to *abandon the dichotomy of matriarchal-patriarchal* to replace it by *egalitarian-dominator*, and thereby avoiding endless discussions if or not in matriarchal cultures males were oppressed by females.

The question in fact is not who dominates whom, *but if a given culture runs on a dominator paradigm or on an egalitarian paradigm*. It is now shared by the majority of scientists that what we formerly called 'matriarchal cultures' were more egalitarian than the subsequent patriarchal or dominator culture. Thus, a way back to love obviously will have to consider a sort of *Archaic Revival*, to speak with Terence McKenna.

—Terence McKenna, The Archaic Revival (1992).

Riane Eisler writes in *Sacred Pleasure (1996), p. 294:*

> This notion that man can, and should, have absolute dominion over the 'chaotic' powers of nature and woman (both of which are in Babylonian legend symbolized by the goddess Tiamat) is what ultimately lies behind man's famous 'conquest of nature'—a conquest that is today puncturing holes in the earth's ozone layer, destroying our forests, polluting our air and water, and increasingly threatening the welfare, and even survival, of thousands of living species, including our own. This is also what lies behind a medical approach to the human body that all too often relies on unnecessary and/or harmful chemical and surgical intrusions – an approach that in Western medicine goes back to the 'heroic' remedies developed by the Church-trained doctors who during the late Middle Ages gradually replaced traditional healers (many of them women burned as witches) and their more natural herbal and other treatments. For here too the guiding philosophy is one of omniscient doctors giving orders and of detached external control; in short, of domination over rather than partnership with nature.

The Early Natural Order

Ancient pre-patriarchal societies really stand out in that they did not practice dominance of females over males but a form of *male-female partnership* that so far was not established again in human society.

This represents a painful loss in wistfulness and balance for the entire globe, and is part of our today's conflictual position with nature, and the fact that we have destroyed most of the globe's ecosphere to a point of no-return.

The Nonviolent Trobriands

As early as in 1929, Malinowski published his report on the sexual life of the Trobriands in which he draws the reader's attention particularly to the sexual life of children and adolescents.

—Bronislaw Malinowski, The Sexual Life of Savages in North West Melanesia (1929) and Sex and Repression in Savage Society (1927).

Malinowski observed, not without surprise, high sexual permissiveness toward children's free sexual play. More generally, he noted the total *absence of a morality* that condemns sexuality in children. Instead, he observed, children engage in free sexual play from early age. Initiatory rites, Malinowski found in addition, were absent with the Trobriands since children were initiated from about three years onwards, generally by older children, in all forms of sexual play. This play is completely nonviolent and includes, with the older children, coitus.

The most interesting finding for Malinowski was that in this culture violence was as good as non-existing and that there were equally as good as no sexual dysfunctions. Trobriands were found to be almost ideal marriage partners and divorce was a rare exception. But what is more, violent crimes were found to be equally non-existent and incest strongly tabooed and inhibited by social norms.

Other researchers found similar phenomena with the Muria tribe in South India where children stay until their maturity in so-called *ghotuls* where they live their sexuality freely and in utter promiscuity. Older children initiate younger ones progressively into sexual play.

—V. Elwin, The Muria and their Ghotul (1947), and Richard Currier, Juvenile Sexuality in Global Perspective (1981), 9 ff.

THE EARLY NATURAL ORDER

These researchers found that after a phase of total promiscuity, the children, from the age of puberty, form and maintain strong bonds and partnerships that are based not on sexual attraction, but on love.

They further found that these first steady relationships formed the basis for later marriages that, regularly, last life-long.

Some researchers and sociologists allege these findings had no significant meaning for our culture since they could not be generalized. However, such arguments assume that man, depending on cultural conditioning, was basically different from one culture to the other. This assumption is questionable, for the biological foundations are with all human beings the same, regardless of cultural, social or religious conditioning. If anthropological or psychological insights were valid only for a given culture, how could psychoanalysis which was founded by Sigmund Freud in Austria be successfully applied in the United States or even in India or South America?

One cannot simply disregard the extensive field studies of highly qualified anthropologists such as Malinowski or Margaret Mead to wipe under the carpet with unscientific publications for the popular press. Political and social conservatism has many faces; if often goes subtle ways in order to vacuum-clean truths that are against the reigning ideology.

> —See, for example, Margaret Mead, Sex and Temperament in Three Primitive Societies (1935).

Interestingly, neither Bronislaw Malinowski nor Margaret Mead have found *pedophilia* present in Melanesia's Trobriand culture where children enjoy the utmost of emotional and sexual freedom. In fact, typically, children in this culture are sexually active with peers, and not with adults.

In other tribal cultures, a bit around the world, pederasty, which is practiced with pubescent boys, has a quite limited function and is a temporary thing to happen. It mainly serves to accompany the boys' initiation in the adult male group. Still more so, men or women being sexually attracted to prepubescent children, refusing to have sexual conduct with adults is something almost non-existent in tribal cultures. This is why I conclude it must be somehow related to specific factors within the cultural setup of dominator civilization. And why not? The question is open. I cannot say I have found a definite answer as to the etiology of pedophilia, while I have found some quite convincing factors that certainly contribute to adults' erotic attraction for minors, and most of these factors are related to our overprotective and often emotionally abusive child-rearing practices combined with the nuclear family structure that does not really allow children to grow away from their mothers and out from the symbolic uterus. They cannot for that reason really build autonomy, which I found is one of the *Eight Dynamic Patterns of Living* in most tribal nations.

But this doesn't mean we should stigmatize pedophilia. Much to the contrary, I am convinced it came up as a social pathology for healing an even greater pathology, something like a social catalyzer for outsourcing childcare from the dysfunctional family toward a new kind of family.

All my research boiled down to nonviolent pedophilia representing not chaos and violence but in the contrary healing and peace, as it acts counter to overprotecting parenting and allows children to engage in erotic friendships outside of the family. It also allows children to project some of their incestuous desires upon adults other than their parents, which was a major pro-pedophilia argument voiced by the late Françoise Dolto in a conversation back in 1986. In that interview I had with her in her Paris residence, the famous child psychoanalyst vehemently pleaded for socially coding adult-child sexual relationships.

I am convinced that all desires that nature creates are purposeful and intelligent. To my knowledge, pedophilia has never been considered under this header, hence the unique stance and importance of my research and the conclusions I draw from them.

Yin-Yang Balance

The primordial energy, when working on the earth plane, manifests in dualistic form, as complementary energies, called *yin and yang*.

It is essential for understanding the natural order that there is a fundamental balance in all of living, and this balance is maintained by the dualistic complementarity, the polarity of *yin* and *yang*. As we will see later in this study, this fundamental balance has been deeply shattered by the destruction of the natural order in form of modern civilization.

THE EARLY NATURAL ORDER

Both of the energetic poles are associated with certain characteristics. However, it would be wrong to identify *yin* with female and *yang* with male. It is not that simplistic. *Yin* can well be associated with the female principle but this does not mean that it is identical with it. It's actually a bit like in the cabalistic system. We talk about corresponding characteristics or elements, and the system as such is one of *corresponding relationships*.

Yin can be said to correspond with water, the female principle, the color black, the direction down or a landscape that is flat. *Yang* can be said to correspond with fire, the male principle, the color white, the direction up or with a landscape that is mountainous.

In every *yin* there is a bit of *yang*, and in every *yang* a bit of *yin*. This bit is the essence that is multiplied once the point of culmination has been passed. What that means is that for example *yin* moves towards its fullness in order to culminate and swap its nature into *yang*. *Yang*, when it culminates, becomes *yin*.

That is why we can say change is programmed into the very essence of the *yin-yang dualism* and thus, change cannot be avoided. We can even go as far as saying that the very fact of *change* is the proof that we deal with a living thing. If there is no change, there is no movement and, as a result, no life. Life is change, living movement.

Pleasure, the Prime Regulator

Early Love for Pleasure

I had my own opinion about pleasure since I was a small child. I had intuitively understood that pleasure is a *vital function of all living* and that I would not allow anybody to deprive me of it. I have grown up in very abusive homes in Germany, where I was beaten almost every day by the perverse female educators, and where we were put in the dark cellar for hours, where babies were just 'forgotten' on the potty, where we had to eat burnt and rotten food, and when we did not want to eat it, we were forced the food down the throat with a spoon, and when we vomited it out, we were fed again with our own vomit.

No, it was not a Nazi concentration camp, but a Catholic children's home back in the 1960s in

THE EARLY NATURAL ORDER

Germany. But despite all this violence, against which my mother never did anything while she knew about it, I *never allowed anybody to interfere with my pleasure function* – and I intuitively knew why. I knew it was vital to have pleasure and that it serves survival, and especially under such conditions.

My mother jokingly said once I was born with a huge erection and I can remember to have masturbated as early as age three. I can remember events until my 2nd year of life, which actually indicates that I have not been traumatized by the brute and violent environment, nor by the beatings, otherwise I would suffer from *childhood amnesia* like most other people. So wow could I *consciously* go through this, without suffering trauma?

Because I had my pleasure, my body pleasure and kept it intact all through my childhood and youth. When I was ten my mother put me in a boarding where I had sexual relations with a peer boy of my age until we both left the school at age eighteen. It was the most happy and fulfilled love relation I ever lived, and our wonderful love helped him and me to overcome the sadistic school environment with all its hidden and open violence both from the side of teachers and peer boys. Three of my school mates had committed suicide, one by taking sleeping pills and two by jumping from high buildings, and for the always same reason: just at the verge of entering university, they had met girls who refused to have sex with them. This is how it comes when you deny pleasure or you wait for the grace of a girl to help you satisfy your longing. Then you wait until your death.

Lovers and Fuckers

And later research confirmed in full that I had practiced the right approach in this matter from earliest age, that I had done right to not let this perverse sadistic violent society pervert me. And I have observed the boys in that boarding, and it was obvious that they formed two groups: those, the minority, who, like me, allowed themselves to have sex and embrace the particular situation in which we were, separated from girls because of stupid moralism, and those, the majority, who suffocated from the emotional and sexual pressure they were building up, the stress of school and the jealousies and fights in the peer group.

THE EARLY NATURAL ORDER

These boys, the majority, we called them the *fuckers*, and us the *lovers*, and their behavior was markedly different from ours. Whereas we were able to solve conflicts peacefully, by communication, by dialogue, they were hard-minded, brute, vulgar, morality-centered, violent and sadistic; whereas we were relaxed and flexible, and kept a sense of humor, they were rigid and stressed, conditioned toward performance, and highly irritable. The slightest word they found displeasing and they would leash out against a smaller boy. After all, it was obvious that *our group was way better adjusted* than they were.

I simply think that this was a sample out of society at large. And they were really suffering from the Oedipal drama or trauma, the unresolved hangup with their *Oedipus Complex*, result of their ruthless conditioning toward *Oedipal Culture*.

One of my sayings during these years of growing up in a dull, perverse and utterly stupid society was:

—Sex is a function of intelligence. Stupid people are not sexual.

I still believe this today, as life showed me many times that it's true. Stupid people are not only not sexual, they also persecute those who are. And this despite the fact that they represent the majority and make the laws. They make down love and sexuality in all their doings, and that is why our sexuality as a whole got more and more perverse since the moment *compulsive morality* was introduced in our culture, which was about from the time of the *Code of Hammurabi*. They have turned love and joyful mating into hate and persecution, and eternal fucking. That's what they are good in, fucking others down, dominating the planet and transform Mother Earth into a trash container. They have fucked down and raped down this beautiful planet to what it is today, a dying planet. And the culture they have created truly is a *rape and murder culture* that has ruthlessly massacred millions of natives around the world, and then, and despite of all this, they have the hubris to go around and preach 'love your neighbor,' these hypocrites!

The Perversion of Pleasure into Violence under Patriarchy

When we know the intricacies of pleasure, we are wiser, and we can begin to understand the Earth. It is significant that in nature-bound cultures, the theme of *sexual lust* is met with humor and with a smiling comprehension that comes from the soul, and from the heart. It's a deep understanding of pleasure and its function in the Web of Life, an intuitive understanding that comes from observing nature, the daily copulation of beings that are alive, in all realms of existence.

And when you come to monotheism-based cultures, with their endless sex taboos, you see that this humor has been transformed into fear. They fear lust more than war, and rape more than murder. They fear all around the pleasure theme, they fear touch and being touched, they fear being fondled, they fear being close to each other.

Thus, simply so, they are perverted from what is natural and good. And the worst is that this perversion, this alienation from nature is *officially recognized and taught as morality*.

They make sure it's firmly rooted in the minds and hearts of children during the slow and gradual soul murder they call 'education'.

NATURAL ORDER

Children know lust very well, simply because they are alive, and that is why they are so violently turned away from it, using all kinds of psychological and physical tortures.

We have been inculcated with life hate, not the love for life, by this idiotic culture and so-called religion – which is of course no religion, but organized stupidity. And when a man naturally expresses lust, and worse even when it's a woman, they are frowned upon, but after so and so many glasses of alcohol during party time, they are given a temporary license –under certain well-defined conditions.

But that license does by no means change the inner setup of it all, the general prohibition, the *general disgust* that is culturally fed and maintained against all that is around the body and the theme of lust. And then we see the cancer rates and the heart disease rates growing in our Western cadaver culture – and we wonder. Why do we wonder? There is nothing to wonder. When you destroy lust, you destroy life, and cancer is virtually a disease where the body, as with the plague, the *pestilence*, rots along and slowly decays under the heavy burden of accumulated morality, accumulated prohibitions,

THE EARLY NATURAL ORDER

accumulated denials of life and of love. But of course, conditioned as they are toward their upside-down worldview, they see not their social and cultural cancer as the plague, but call lust itself a plague, as illustrated by the book title *The Plague of Lust*, by Julius Rosenbaum.

Life requires love and lust, otherwise it dies. That's simply a fact of nature. The *Plague* or *Black Death*, the terrible epidemic that killed millions over several centuries from the Middle Ages to the Renaissance was the result not of bacteria, but of rampant sexual misery. There has never been a plague or anything even remotely similar with nature-loving and sexually active native people, at no time, nowhere in the world.

Pleasure and Intelligence

Herbert James Campbell, a British neurologist, found in twenty-five years of research a universal principle which regulates the whole of our well-being and intelligence, *the pleasure principle*. This sounds like Freud, but it has little to do with psychoanalysis or psychology. What we are facing here are facts proven by natural science, by neurology.

NATURAL ORDER

In 1973, Campbell published his book *The Pleasure Areas*, which represents a summery of many years of neurological research. Campbell succeeded in demonstrating that our entire thinking and living is primarily motivated by pleasure. He found that pleasure manifests not only in a tactile-sensual or sexual way, but also as non-sensual, intellectual or spiritual pleasure. With these findings, the old theoretical controversy if man was primarily a biological or a spiritual being became obsolete, for it is in the first place our striving for pleasure that induces certain interests in us, that drives us to certain actions and that lets us choose certain pathways in life.

Campbell made the revolutionary discovery that our preferences literally change our brain's neuronet. During childhood and depending on the outside stimuli we are exposed to, certain *preferred pathways* are traced in our brain, which means that specific neural connections are established that serve the information flow and the memory storage. The number of those connections is namely an indicator for intelligence.

THE EARLY NATURAL ORDER

The more of those preferred pathways exist in the brain of a person, the more lively appears that person, the more multi-vectorial will the person be in their approach to managing their life, the more interested she will be in in a large variety of disciplines, and the quicker she will achieve integrating new knowledge into existing memory.

High memorization, Campbell found, is depending on how easily new information can be added on to existing pathways of information. Logically, the more of those pathways exist, the better! Many preferred pathways make for high flexibility and the capacity to adapt easily to new circumstances.

And it goes without saying that sexual experience and variance in sexual relationships makes for many preferred pathways to be established, especially in childhood and adolescence. I would go as far as saying that sexuality is a primary means, and an especially effective way to establish preferred pathways in the brain and thus to raise intelligence. So this research fully confirms my early intuition.

PLEASURE AND TOUCH

This is true not only for full-range penetrative sexuality, but also and with special significance for *tactile sexuality* and non-sexual tactile contact, skin-skin contact among adults and children, and cross-generationally the mutually desired *peau-à-peau* between parents and children, tutelary and non-tutelary adults and children, adults and adolescents, as well as adolescents and children.

Campbell's research indicates that the repression of pleasure that is since centuries rampant in our Judeo-Christian societies has negatively infringed upon human evolution and impaired the integrity of our psychosomatic health. This is exactly what Wilhelm Reich found – without having had at his disposition Campbell's neurological findings.

Not only neurologists have thought about the basic functions of life and living, but also people who were formerly active in totally different fields of science. American scientists Ashley Montagu and James W. Prescott had different points of departure for their extensive research. Montagu and Harlow wanted to know why small rhesus apes died when they were deprived from their mother while they survived when a soft cloth doll was put in the cage as surrogate of motherly tactile affection. Prescott researched on the origins of violence. He did from the start oppose the age-old myth that man was per se a violent creature even though human history, or what historians saw of it, seemed to prove it.

These and other scientists basically concluded that tactile stimulation of the infant is a main source of early pleasure gratification and a condition for human health, for harmony, and for peace. Ashley Montagu's research developed quickly a specific focus on the human skin as a primary pleasure provider. Grant's *Method of Anatomy* defines the skin as the most extended and the most varied of all our sensory organs.

Montagu's study *Touching: The Human Significance of the Skin (1978)* was the final result of thirty years of skin research, not only Montagu's, but of many other scientists whose research Montagu presents and evaluates in his study.

Ashley Montagu's research is of paramount importance for our *understanding of tactile stimulation in early childhood*. His specific research focus was upon the mammal mothers' licking their young. He found most zoologists are conscious of the importance of motherly licking for the survival of the offspring. He discovered that it is first of all the *perineal zone* of the young that the mother preferably and repeatedly licks. Experiments in which mammal mothers were impeded from licking this zone of the young resulted in functional disturbances or even chronic sickness of the genito-urinary tract of the young animals.

Ashley Montagu concluded that the licking does not serve hygienic purposes only, but is intended to provide a tactile stimulation for the organs underlying the part of the skin that is licked. However, Montagu further concluded, licking is exceptional to happen in the mother-child relationship with primates or humans.

Most researchers found that for humans, licking was gradually replaced by *eye or skin contact between mother and child.* The tactile needs of the small child seem to correspond to the desire of the parents to express love through tactile affection such as kissing or fondling, or pressing the child's naked body against one's own during sleep or rest, which is common with Eskimos and most other native tribes.

THE EARLY NATURAL ORDER

In the run of industrial civilization, however, this has changed fundamentally. 1960s pediatrics and child psychologists recommended parents to put their children in separate rooms and beds so that parents and children were physically separated. This is the main reason why the civilized child gets much less tactile stimulation in early childhood than children from tribal cultures, a fact that Jean Liedloff demonstrated in her alarming book *The Continuum Concept* which was first published in 1977.

Ashley Montagu and James W. Prescott, coming from different scientific angles, agree that early tactile stimulation is paramount for the psychic and physical health of the child and later adult. A direct relationship was discovered by both researchers between early tactile stimulation and the functioning of the immune system of the child.

This relationship was corroborated by France's world-famous obstetricians, Frederick Leboyer and Michel Odent. As Michel Odent writes in his book *La Santé Primale*:

> It is not yet completely understood that sensorial perceptions at the beginning of life can be a way to stimulate the 'primary brain,' at a time when the 'system

of primary adaptation' is not yet grown to maturity. More specifically, this signifies for example that, if one fondles a human baby or an animal baby, one also stimulates his immune system. (Id., p. 24, Translation mine)

Montagu states that love was once defined as the *harmony of two souls and the contact of two epidermises.* In this sense the *peau à peau* that is nowadays even recommended by pediatricians, is a foremost condition for the healthy growth of children, the good functioning of their immune system and, last not least, the early creation of *preferred pathways* in their brains. Skin contact thus favors high intelligence.

James W. Prescott's research particularly focused on the consequences of early tactile deprivation in the form of shortened or lacking breastfeeding. In his article *Body Pleasure and the Origins of Violence* Prescott uses R.B. Textor's supra-cultural statistics to scientifically corroborate his highly far-reaching and politically relevant conclusions.

>—Bulletin of the Atomic Scientists, 10-20 (1975), partly reprinted in: The Futurist, April, 1975.

Already in the 1930s Wilhelm Reich disproved the very widespread misconception that sadistic and destructive tendencies were part of human nature. He strongly opposed Freud and his theory of a *death instinct*, arguing that destructive instincts are but secondary drives, a direct consequence of the cultural repression of the natural sexual instinct which resulted in collective neurosis. In his book *Children of the Future (1950/1984)*, he outlines an emotionally and psychosexually sane education of children for a society that accepts biogenic regulation, the natural self-regulation of biosystems.

Pleasure and Violence

Reich's findings, at the time violently opposed by the majority of his scientific colleagues, are confirmed by Prescott's findings which bring statistic evidence as to the malleability of the human individual through his early tactile experiences or the absence of such experiences:

> Recent research supports the point of view that the deprivation of physical pleasure is a major ingredient in the expression of physical violence. The common association of sex with violence provides a clue to understanding physical violence in terms of deprivation of physical pleasure. (…) Although physical pleasure and physical violence seem worlds apart, there seems to be a subtle and intimate connection between the two. Until the relationship between pleasure and violence is

understood, violence will continue to escalate. (Id., pp. 10-11)

Unless the causes of violence are isolated and treated, we will continue to live in a world of fear and apprehension. Unfortunately, violence is often offered as a solution to violence. Many law enforcement officials advocate 'get tough' policies as the best method to reduce crime. Imprisoning people, our usual way of dealing with crime, will not solve the problem, because the causes of violence lie in our basic values and the way in which we bring up our children and youth. Physical punishment, violent films and TV programs teach our children that physical violence is normal. (Id., p. 10)

Prescott thus fully confirmed Reich's earlier research and corroborated his socio-economic and sex-economic findings. More specifically, James W. Prescott found a noteworthy relationship between pleasure and violence. Referring to laboratory experiments with animals, he could detect a sort of *reciprocal relationship* between pleasure and violence, i.e. that the presence of pleasure inhibits violence—and vice versa. Prescott states:

> A raging, violent animal will abruptly calm down when electrodes stimulate the pleasure centers of its brain. Likewise, stimulating the violence centers in the brain can terminate the animal's sensual pleasure and peaceful behavior. When the brain's pleasure circuits are 'on' the violence circuits are 'off,' and vice versa. Among human beings, a pleasure-prone personality rarely

displays violence or aggressive behaviors, and a violent personality has little ability to tolerate, experience, or enjoy sensuously pleasing activities. As either violence or pleasure goes up, the other goes down. (Id.)

Furthermore, Prescott found a direct relationship between the child rearing methods of a given culture, and the *level of violence* that reigns in that culture. In detail, he found that societies that tend to rear children in a rather Spartan way, hostile to pleasure and with little or no tactile affection cherish in their value system various forms of violence, they do warfare, torture their enemies, practice slavery and progeny and concede to women and children a rather low social status; these societies also exhibit a high crime rate.

Another violence-indicating parameter in a society, Prescott found, is physical violence toward children in form of corporal punishment. Furthermore, repression or tolerance of children's sexual life plays a decisive role in the assessment if a society has a high or low violence potential:

> Thus, we seem to have a firmly based principle: Physically affectionate human societies are highly unlikely to be physically violent. Accordingly, when physical affection and pleasure during adolescence as well as infancy are related to measures of violence, we

find direct evidence of a significant relationship between the punishment of premarital sex behaviors and various measures of crime and violence. (Id., p. 13)

As a result of his research, Prescott advocates the total abolishment of corporal punishment of children, a rise of the social status of women, extended breastfeeding (2.5 years and longer), baby carrying, abundant tactile contact between parents and children, the reinstitution of the extended family, the reintegration of the elder and a more active participation of men with child-rearing and the granting of physical affection to children in their role as fathers or educators.

—See James W. Prescott, Deprivation of Physical Affection as a Primary Process in the Development of Physical Violence (1979), pp. 77, 78.

Child Sexuality and its Detractors

Children and Sex

I began in the 1980s to research on child sexuality, at the same time I started research on *parent-child co-dependence* and *emotional abuse* of children, with the ultimate goal to find ways out of the dilemma in the form of new educational and social policies. This research showed me that our society is strangely fixated upon, and almost obsessed about, child abuse, instead of looking what was first, the hen or the egg? In truth, what was first is love, and natural sensuality that at times, and under certain conditions does become sexual.

That sounds like a commonplace but is not. We have the same problem with Western medicine that is fixated upon the pathological but that never has defined *what health is actually about*, what exactly health looks like—obviously so, health is more than the absence of illness. So it is with child sexuality.

When parents contend that their child be never sexual and thus that there was a total absence of child sexual behavior, this doesn't necessarily mean that we are dealing with a healthy family setup; it rather indicates that these parents' perception of their child is veiled, obstructed, biased, because of their own uneasiness with being sexual in the first place.

Hence, the fact that children are sexual is more than just the absence of their 'not being sexual;' by the same token, and contrary to what Sigmund Freud and the majority of child psychologists assert, child sexuality is *more than auto-eroticism*, more than masturbation, and more than talking freely about sex. It is the *faculty to love*, and to lovingly embrace mates other than the parents, for the parents are not ideal love mates for the child. To assume children have to choose their own parents as their first love mates programs for individual and social neurosis, and here Freud has done us a disservice with his construct of the *Oedipus Complex*.

Children grow into healthy sexual beings not because of mating with their parents, even if such mating is supposed to be only on the fantasy level, but *because they lovingly embrace people they fall in love with*, people from outside of the family, peers or adults other than their parents.

This is simply so, if our puritanical and hypocrite society allow or affirms it or not, for, to be true, here is the stumble stone called *pedophilia*, again and again advanced as the torture agenda and the invasion of chaotic liberalism. No, if a child chooses love with an adult, this is *not* a case for pedophilia, but a *love choice*, and society has to respect it, for otherwise we are not talking about liberal education, we are not talking about human rights and democracy, but upon censorship, and tyranny. How erotic is a child supposed to be, or not supposed to be, when it goes to actual love-making? If we take mechanical sex education serious, we can forget about love for it has no place in a machine. If the nation learnt by such kind of sex education, we know why the nation is perverse sexually. Ever thought no sex education is needed in a society that really respects children?

A Note on Child Trauma

As Anna Freud's research in British war shelters and nurseries during World War II, showed, children are not easily put in a condition of fear, constant anxiety or even trauma.

—See Anna Freud, War and Children (1943).

For this to happen, something must have shattered their emotions, or their emotional relationships with any person they love. It is significant to see that children are not per se fearful or afraid of anything; they are not scared of the facts of life, even if there is danger. Children are not that easily traumatized, except they have been educated to be anxious, which is typically the case when they are raised by neurotic parents. We can thus conclude that children are not afraid of sexuality or sexual encounters of any kind, of course, provided they are not outright violated or abducted or otherwise treated with other disrespect; but they would not be afraid of having pleasurable sex with a person they love, peer or adult. Now, when you see that this is the natural position, you will also understand that when you see a child being highly anxious about sex or nudity, *something has happened that was not okay*, or the child is raised by parents who are not comfortable with their own sexuality, who are full of shame and life-denial, and who are unhappy in their couple relation.

Adolescents and Sex

Many parents may find it funny or scurrilous when their small child shows signs of sexual arousal or curiosity. They may be naturally permissive in the face of it all, as it's more or less something that is confined to the home and where little will be known outside. But these same parents may react with bewilderment to the budding sexuality of their adolescent boy or girl, fearing basically two things:

THE EARLY NATURAL ORDER

—that the adolescent may stay out for parts of the night or whole nights to pass in the house of their friend or friends;

—that their adolescent may talk about their desire with peers or children of the neighborhood who then find out that the parents practice a sort of permissive education; what they fear is to be marginalized by their neighbors or friends who may have lesser permissive attitudes in this respect. The basic problem is that many parents are uncomfortable within the couple, regarding their own long-term experience of intimacy.

What is the problem with being more permissive than a neighbor? We are all different. In one single small neighborhood, you may find a miniature model of all possible behavior modes in society as a whole.

As this is so, parents who are healthy and comfortable in the couple will not resent that others may find fault with their particular child rearing paradigm, and they will know how to defend their position. When they worry about what others say, they have not really found their own position and probably would fare better to be lesser permissive, but comfortable with their decisions.

NATURAL ORDER

When parents are afraid of their adolescent children sleeping in the house of a friend, this can have various reasons. They may first want to check out these people, which is normal and responsible behavior, but if they find them okay and still won't allow their adolescent to sleep there once in a while, then they are probably co-dependent with their adolescent child. I am not talking about a fancy here; co-dependence is something very real. The symptoms are not easy to be dealt with; for example a clear symptom is that the parents once allow the child to sleep with their friend, but do not close an eye for the whole night.

The next morning then, when their child returns, they make long faces and hold a long discourse about *how much they suffered* from the child's absence, that they couldn't sleep the whole night, and that all this proves 'how much they love the child,' and that they would just suffer too much if the child wants this to happen again—and that, *conclusio ad infinitum*, they are against repeating the experience. No wonder then to see that adolescent shocked and traumatized for becoming emotionally entangled with the unresolved issues of his or her parents. And the

THE EARLY NATURAL ORDER

benefits of that night sleeping outside, potentially conducive to the adolescent's building a greater circle of autonomy, are annihilated!

This is how it should not, and unfortunately the majority opinion in our society tends to give right to such perverse parental behavior which leads to pathological emotional entanglement, and which in my view is *emotional child abuse*, instead of giving right to children in their important quest to build autonomy and to grow out from childhood.

In fact, why child sexuality, and all sexuality, is so important is that it helps building personal identity and acts counter to semi-incestuous and general emotional entanglement that is unhealthy *because it tends to inhibit the natural growth of children into responsible and self-reliant adults.*

I contend that without learning the loving embrace, a child or adolescent cannot become a self-reliant adult later on, a person with high self-love and a stable sense of identity.

Our society has as yet a long way to go to affirm this truth, and to give children and adolescents the freedom they need to become persons in their own right, and not pleasing night pillows and tear-catchers for their parents and caretakers.

No Sex Education Needed

Wilhelm Reich and Françoise Dolto coincided in their views that sex education 'always comes too late.' They meant that sex education is always more or less an intellectual circus that doesn't reach the child's unconscious and therefore rests at the surface of the personality. If parents have failed to be permissive enough to allow their children to be sexual, *all sex education can't alter this fact*, and thus will make it only worse.

This was demonstrated some years ago by a large research conducted in nurseries and schools in the San Diego Bay Area; the research results led to serious doubts as to the effectiveness of school-based sex education.

In fact, what was found was that most children suffered from harmless to severe misconceptions about sexuality and procreation in ways that sounded almost absurd. They showed clear signs of confusion, and it became evident that they were not able to integrate the factual knowledge they had been taught

THE EARLY NATURAL ORDER

with their own emotional and imaginal world. This means they have been traumatized by such education, which can surely not be the goal of it.

In my personal view, sex education is pure fake, a guilt reaction that society shows in the face of its own hypocrite and violent denial of children's emotional and sexual needs. It's a palliative medicine that as most of Western medicine is simply useless, crap and fake, a money making machine, a worldwide business, and a science that serves to manipulate public opinion in favor of upholding the old stereotypes and blind spots, so that parents can go on with their denial attitudes and their horrid persecution of children's intimacy with the argument:

—Oh, I can relax about all this, I don't need to know anything about it. School will take care of all this, anyway …

School won't, and such an attitude is simply irresponsible and criminal. What children need is not to copulate emotionally with their parents, but to copulate emotionally and sexually with people they choose as their lovers or mates and who are not part of the family or clan.

Child Sexual Abuse

Child Sexual Abuse as a modern formula for criminalizing child sexuality. It is really the ultimate fake cause. As child sexuality is not conforming to the rules of consumer society, a cause was created that declares children to be *sex offenders* when they engage in the most natural of all games, sexual play with other children.

The official rhetoric is that children live their childhood with a promise, the promise of later integration, of later life, of later pleasures, of later responsibilities, of later freedom. In the meantime they have to play and, first of all, shut up and restrain from criticizing the protection system that cares for their best and that holds them back forcefully from any experience of real life – because that could make them rebellious and thus subversive. It could namely wake them up from the mass hypnosis they are subjected to by their oppressors. Stevi Jackson, in *Childhood and Sexuality (1982)*, notes:

> So sharply are the distinctions between adult and child drawn that the two seem almost to belong to different species: adults are independent, children dependent, adults productive, children non-productive, adults work, children play, adults are involved in the serious business

of life, childhood is supposed to be fun. It is not simply that children are treated as people who have yet to learn the skills and conventions of adult life, but that they are regarded as beings of a different order with needs quite apart from those of the rest of the community. (Id., p. 24)

Pedoemotions and Pedoerotics

Definition

Pedoemotions are temporary, transient, recurring or exclusive emosexual desires and fantasies involving children.

> —I have created the term 'emosexuality' in order to emphasize that sexual orientation is based upon emotional predilection and not vice versa, upon so-called sexual drives, as sexology assumes. The term is not simply a composition of the words emotionality and sexuality. It does not only express that emotions and sexuality naturally swing together, but that this union creates the unique experience that we call love.

While pedoemotions are not primarily sexual, they focus our emotional attention upon children in a way that children become more important, more attractive, more interesting to be with, more captivating and more seducing than they are for a control person with a lesser degree of pedoemotions. Pedoemotions are present in both men and women and their love objects can be either male or female children or in a bisexual form both boys and girls.

What is the Nature of our Emotions?

Emotions are manifestations of the *human energy field*. They are to be found as *emonic vibrations* in the cell plasma. In this sense, emotions are functional and they serve a specific purpose. Pedoemotions manifest in adult men and women so as to ensure our loving care for the young. Without pedoemotions the human race would since long have ceased to exist.

It is important to understand that pedoemotions are *emotions* and not per se sexualized. Why pedoemotions become sexualized in one case and not the other, we do not know yet.

Sexology has so far not found real answers to the question why certain adults make emotional choices for children rather than for adults; this may in part be due to its mechanistic base paradigm. I doubt that the answer can be found with *phallometric assessment* (Freund, Finkelhor, Rind and Bauser et al.); first we are not even talking about sexual response, but about a purely *emotional predilection* for children that may, or not, be sexualized. It has to be seen that assessing pedophilia unfortunately in modern law enforcement is being used as a standard for discarding out and silencing a certain number of intellectuals modern society doesn't like to confront because they are asking pertinent questions about the validity of the present *shut-up-and-better-be-violent* paradigm.

We can find the key to the rationale of pedoerotic attraction in all its forms only from the moment we ask the question why certain people at certain times in their lives make a more or less informed choice that gives an emotional preference for children over adults?

I admit that the answer to this question is disturbing for postmodern international consumer culture because *the consumer child cannot be a sexual child*, and because the answer would reveal that a permissive attitude toward adult-child emotional and sexual attraction leads to lesser violence not only against children, but generally a lesser violence potential within society as a whole.

NATURAL ORDER

What are Pedoemotions?

It is important to realize that pedoemotions do not per se become sexualized. Pedoemotions are positive manifestations of the love for the young and they are connected to the pleasure function. Natural and non-neurotic adults of both sexes enjoy tactile closeness with dressed or undressed children, co-sleeping with children, massaging children or otherwise *bestowing abundant tactile stimulation and care* upon children without being driven, by so doing, to sexually interact with the child.

However, our society more or less silently assumes that the latter will be the case, and that is one of the reasons tactile stimulation of children, and tactile interaction with children, has been turned down in all modern consumer societies; while anthropology shows with much evidence that this abundant tactile care is being practiced in almost all native cultures around the globe. It's also one of the reasons for males to have been discarded almost completely from early childcare in most Western nations.

The difference between an erotically relaxed parent or caretaker of a child and a neurotic and fixated pedophile is that the first is not driven by an

THE EARLY NATURAL ORDER

urge to sexually engage with the child, but enjoys the pleasure of tactile closeness in a relaxed mood and for sharing wellbeing with the child.

Now, and contrary to most mainstream research on pedophilia, I do not assume that all natural childcare is per se non-erotic and selfless, but I affirm the erotic dimension in child care as a strong and natural ingredient of healthy child care.

What is Pedoerotics?

Pedoerotics is the eroticism that develops *when pedoemotions are sexualized.* It is an often blissful situation with corresponding ecstatic feelings and deep moments of euphoria in case the subject embraces the attraction. On the other hand, it is felt as oppressive, as an urge or desire that is difficult to control, when it is repressed.

Regularly, when men or women abuse of children sexually in the sense to gain orgasmic satisfaction without the consent of the child, the pedoerotic desire was repressed, and not embraced. That is why it is important to *liberate and socially code pedoerotics* as a blissful state that sees in children their intrinsic beauty and vibrant eroticism. There is a clear difference between being neurotically driven to sexually conquer a naked child in a need for sexual union, on one hand, and the sharing of tactile pleasure between an adult caretaker and a child, on the other. It is not easy to express this difference in words, but a sensitive, emotionally sane and intelligent human will understand intuitively what I want to convey.

In certain cases, pedoemotions can become sexualized, which may result in the feeling of erotic attraction of an adult toward a child, or an adolescent toward a younger child or baby. Such attraction may, or not, be acted upon. Research has brought to daylight that, however, in most cases, such attraction is not acted out by actual sexual penetration of the child, but more often than not by fondling, kissing, caressing, licking and shared nakedness as well as masturbatory tenderness and sharing.

What is Adult-Child Sex Like?

Actually, from a body language point of view, these interactions are more closely related to a mammal licking their young than to human intercourse. As a matter of fact, as with nurturant care, these interactions tend to bestow upon the child a certain level of tactile affection that the baby, child or adolescent can only profit from. This shows that, contrary to the mainstream tenor on the matter, there is well a loving-and-caring aspect in nonviolent pedophilia!

In addition, it is important to behold that a person, male or female, who occasionally or once in a while experiences a sexualized pedoemotive attraction toward a baby, child or adolescent, or who experiences such attraction over a certain period of time, is *not* a pedophile, because the assessment of pedophilia in my view requires an *exclusiveness of sexualized pedoemotive attraction*; this means that the erotic desire has been condensed into a single focus kind of attraction and corresponding emotional predilection. It's typically the person for whom children tend to be everything in life and who in addition finds that sexual interaction with children, even when there is no penetration or intercourse, is the most important and fulfilling, if not the only variant in their erotic life. In addition there is a divider between violent and narcissistic pedophilia and erotic and abundantly tactile child care. This difference has been voiced even by some organizations of pedophiles who firmly contradict the mainstream assumption that labels them as child rapists and who make up the divider for their own unwritten rules of conduct.

They argue that profiting sexually from a child is a selfish and ego-driven behavior and as such does not qualify as childlove or pedophilia, but as child rape, but that there is a form of caring and non-violent childlove that stands out as a form of caring love in that it is child-centered, and not self-centered. In fact, it is true that erotic attraction to a child and having loving and caring feelings for the child, and actually bestowing such care upon a child, are not two mutually exclusive forms of behavior, as this is said in the mainstream rhetoric. They can well go along with each other, and mutually enrich each other.

It also has to be seen that actually harboring erotic feelings for a child does not per se imply that the person acts out on such feelings. Caretakers, teachers, and parents, may at times fantasize erotically about the children they are taking care of but in the regular case, they may not reveal their feelings to the child. They may experience

THE EARLY NATURAL ORDER

bewilderment, or they may accept those feelings without worry, but one thing is certain: the fact that those feelings are temporary or transient means that those adults are not to be qualified as pedophiles.

I think it is quite important to stress this as a large part of the anxiety and the bewilderment in our society about these issues comes from the labels that society so graciously inflicts upon those who are in any way different from the herd of heterosexual monkeys. When people are labeled according to their behavior, we are not far from fascism. Human beings are not characterized by their behavior because of the simple fact that they can change their behavior, any time, any place and for any reason, including for no reason.

Fact is that all people at times experience erotic attraction to humans younger than themselves, which is simply so, if this truth fits in the mainstream labels is not my business! There are two ways of apprehending reality, the mainstream one and the intelligent one.

References

Lauretta Bender & Abram Blau, Abram
The Reaction of Children to Sexual Relations with Adults

American J. Orthopsychiatry 7 (1937), 500-518

Brant & Tisza
The Sexually Misused Child
American J. Orthopsychiatry, 47(1)(1977)

CHAPTER TWO

The Destruction of the Natural Order

Cartesian Reductionism

The *Cartesian reductionist* science paradigm formulated mainly by René Descartes, Baron d'Holbach and La Mettrie, three influential French philosophers, prepared the ground for a major schizoid split between science and religion, thereby annihilating for the last four hundred years the progress that was made by perennial science thousands of years before.

The Cartesian or Newtonian worldview is a life philosophy marked by a *hypertrophy of deductive and logical thinking* to the detriment of the qualities of the right brain such as associative and imaginative thinking, and generally fantasy. It's also a worldview that generally tends to disregard or deny dreams and dreaming, extrasensorial perception and ESP faculties as well as genuine spirituality. The term *Cartesian* has been coined to mark a similarity in reasoning of Cartesian-minded people with the reductionist philosophical theories of the French mathematician and philosopher René Descartes (1596-1650).

I refer to Cartesian thinking or the Cartesian worldview to demonstrate to which extent Western philosophy and culture is essentially a psychological blind-out of the holistic and organic web that life represents in reality, having split mind and matter into opposite poles.

Historically speaking, it was not Descartes who first construed this schizoid worldview, but the so-called *Eleatic School*, a philosophical movement in ancient Greece that opposed the holistic and organic worldview represented by *Heraclites*; but it was through the Cartesian affirmation and pseudo-scientific corroboration of the ancient Eleatic dualism that in the history of Western science, the left-brained *reductionist* approach to reality, which is actually a fallacy of perception, became the dominant science paradigm. Fritjof Capra, in his bestselling book *The Tao of Physics (1975/2000)*, observes:

THE DESTRUCTION OF THE NATURAL ORDER

> The birth of modern science was preceded and accompanied by a development of philosophical thought which led to an extreme formulation of the spirit/matter dualism. This formulation appeared in the seventeenth century in the philosophy of René Descartes who based his view of nature on a fundamental division into two separate and independent realms: that of mind (res cogitans), and that of matter (res extensa). The 'Cartesian' division allowed scientists to treat matter as dead and completely separate from themselves, and to see the material world as a multitude of different objects assembled into a huge machine. (Id., p. 8)

At the same time, this worldview became the dominator doctrine in child education, and the systematic emotional castration and desexualization of the consumer child became the main concern and a constant feature in modern Western child rearing.

And it is exactly this reductionist and nature-hostile worldview that has established the foundation for the *Oedipal Culture* or postmodern international consumer culture that is the present-day apocalyptic vintage of cultural schizophrenia, abysmally infatuated as it is with a set of paranoid, persecutory and highly violent mainstream values that it uses to hypnotize consumers into the Brave New World of total consumption.

And yet presently, in the first decade of the 21st century, even mainstream science gurus such as Fritjof Capra declare Cartesianism to be overruled by the new physics and the emerging holistic sciences that are presently breaking through as a preparation for a completely new worldview in the West, while in Eastern culture this organic, holistic worldview was always the prevailing one. Reminding us of Einstein's genius that was never affected by the reductionism of the schizoid Cartesian worldview, Capra observes:

> Einstein strongly believed in nature's inherent harmony, and his deepest concern throughout his scientific life was to find a unified foundation of physics. He began to move toward his goal by constructing a common framework for electrodynamics and mechanics, the two separate theories of classical physics. This framework is known as the special theory of relativity. It unified and completed the structure of classical physics, but at the same time it involved drastic changes in the traditional concepts of space and time and undermined one of the foundations of the Newtonian world view. (Id., p. 50)

With the emergence of quantum physics, at the beginning of the 20th century, the Newtonian worldview began to break apart, and so did the values connected to it. It was particularly the nature of the light that was stirring the controversy that ultimately inflicted the death blow to Cartesian science. The

THE DESTRUCTION OF THE NATURAL ORDER

events that led there were historical and dramatic. Fritjof Capra remembers:

> The whole development started when Max Planck discovered that the energy of heat radiation is not emitted continuously, but appears in the form of 'energy packets'. Einstein called these energy packets 'quanta' and recognized them as a fundamental aspect of nature. He was bold enough to postulate that light and every other form of electromagnetic radiation can appear not only as electromagnetic waves, but also in the form of these quanta. The light quanta, which gave quantum theory its name, have since been accepted as bona fide particles of a special kind, however, massless and always traveling with the speed of light. ... At the subatomic level, matter does not exist with certainty at definite places, but rather shows 'tendencies to exist', and atomic events do not occur with certainty at definite times and in definite ways, but rather show 'tendencies to occur'. In the formalism of quantum theory these tendencies are expressed as probabilities and are associated with mathematical quantities which take the form of waves. This is why particles can be waves at the same time. (Id., p. 56)

Quantum physics really has demolished the classical Newtonian worldview with its strict determinism. As Fritjof Capra concludes, a careful observation of subatomic particles shows that the observation of these particles gives meaning only when they are seen not as isolated entities, but when understood as *interconnections between the preparation of an experiment and the subsequent measurement.* Quantum physics reveals a basic oneness of the universe at least at a subatomic level of observation, which is exactly what perennial science and mystical traditions of the East and West always have assumed as the main characteristic of reality.

In his second bestselling book, *The Turning Point (1987)*, Fritjof Capra then concludes this insight and extrapolates it beyond the realm of physics:

> In contrast to the mechanistic Cartesian view of the world, the world view emerging from modern physics can be characterized by worlds like organic, holistic, and ecological, It might also be called a systems view, in the sense of general systems theory. The universe is no longer seen as a machine, made up of a multitude of objects, but has to be pictured as one indivisible dynamic whole whose parts are essentially interrelated and can be understood only as patterns of a cosmic process. (Id., p. 66)

When we observe the current science revolution and acknowledge its growing impact upon the emerging global worldview and consciousness, we are forced to evaluate the current tendencies to deny complexity and reaffirm age-old stereotypes of control and survival as a *psychological resistance*

THE DESTRUCTION OF THE NATURAL ORDER

against the inevitable revolution of consciousness that is virtually at our door step. It is here where we, as scientists, poets, philosophers and educational professionals must step in and prepare the world around us for the shift of perspective the mainstream media are carefully hiding behind their unending murder-and-abuse spectacles.

This is actually another vintage of reductionism put on stage, this time in a modern costume, but it exemplifies what I am saying in this book. And it shows that in most cases, reductionism is a fear-reaction to the natural changes in the living that occur virtually every day. When I am afraid of change, and of the complexity of life, I try to use my intellect to reduce this very complexity and unpredictability of life to a concept that I can control, or that I *think* I can control.

Reductionism is a typical modern-day phenomenon. It is unthinkable to see it placed in the Middle-Ages, for example, and even in the Renaissance. Historically, and not surprisingly so, it coincided with the formulation of Cartesianism; it has taken root with the French philosophers René Descartes (1596-1650) and La Mettrie (1709-1751) who were considering humans as machines and nature as a complex yet mechanical machinery. Cartesianism at its very root is but reductionism, perhaps a sophisticated vintage of it, but still; what it does is to reduce life, and science, to a kind of minimalist concept that oversimplifies the complexity inherent in living processes, and paints life as a clockwork, a machine, a robotic scheme, basically denying life its quality to be, first of all, a highly complex structure of *total information*. Usually, reductionism is defined as 'reducing the nature of complex things to the nature of sums of simpler or more fundamental things.' This can be said of objects, phenomena, explanations, theories, and meanings. However when reductionism meets Cartesianism, the result is much more devastating because the outcome of this paradigm is not just a denial of complexity, but a denial of life, of its organic and self-perpetuating nature, its intrinsic intelligence, which in turn is a result of its total information network.

We can observe that with a holistic view of the universe as it is part of the New Age, the mechanical reductionism of Darwinian evolutionary psychology is overcome and left behind, and science presently

THE DESTRUCTION OF THE NATURAL ORDER

changes many of its fundamental assumptions and paradigms because of this shift in understanding nature, human nature and the cosmos at large.

Let me give a few typical examples for reductionism in scientific texts and popular imaging. For example, it is written by Rupert Sheldrake in his book *A New Science of Life (1995)* that the old idea of a cosmic life energy, life force or vital energy was but a *vitalistic theory*. What Sheldrake says throughout his book is that there is no such cosmic life energy, and he thus was reducing the whole idea of a cosmic energy to the term *vitalism*—which is an intellectual concept made up by skeptics.

It has to be seen that often in science and also in political scripts and writings, reductionism is used for belittling, or outright downplaying important concepts and phenomena of life, thereby manipulating scientific evaluation and sometimes also, for influencing public opinion. We need to simply acknowledge that Sheldrake, when reducing the energy flow or information network of living systems to the intellectual concept of vitalism, was *unscientific*, for he did with not one sentence discuss the matter in substance.

NATURAL ORDER

This is against the scientific method, and typically when that happens we know the person is defensive with regard to certain insights he or she may obtain when opening the door to *real understanding*. Sheldrake, while he would certainly deny this label, betrays in so far that he is an ideologist, not a scientist. He is afraid that if he acknowledges the perennial concept of the cosmic life energy, he would mess up the 'purity' of his own concept of morphic resonance, which is after all the same thing in a different wrapper.

Morphic resonance basically says the same what a growing number of scientists say about the total information field, or *quantum vacuum*, at the basis of all life. It says that there is such an information field that totally connects all the vectors built in the moving matter of life. If one may call this phenomenon cosmic life energy, field, vacuum, resonance or otherwise is of strictly secondary importance.

The next example, as obvious as it appears, plays a predominant role in our daily images, the images we are bombarded with, virtually since childhood, in our mass media. For example, in magazines for pleasure, sex and lifestyle women are depicted as mere bodies, or objects. What you get to see, in the worst case, is a pair of breasts, a belly button and a vagina, with no connected head, legs and feet.

Such kind of photos thus reduce women or girls to their erotic attributes and organs, thereby subtly suggesting that 'the rest is of minor importance;' from there the associative pathway is open for derivative assumptions such as 'they anyway do not have a brain' or 'females are anyway best suited as pleasure toys' or 'they're good for the one and only thing, and for bearing children, but otherwise pretty useless.'

Thus, when you carefully observe the *associative and suggestive* qualities of reductionism in photography and film, you become perhaps aware of the really destructive effects of being reductionist when it is regarding life, and human beings.

The Upsurge of Moralism

Moralism is a short term for a huge dilemma. It has nothing to do with genuine morality; in fact, moralism is a perversion of true morality. One of the first perpetrators of violent moralism in human history was the Babylonian King *Hammurabi*. He was also the first ruler who used moralism as a political strategy.

Moralism is a cover paradigm and fake concern when there is in reality the most cynical indifference both in society and in government, and where there is a *high level of structural and domestic violence* and a strong suppression of truth and free speech. Every form of political fascism begins where these basic conditions are met; moralism is used strategically for the following pursuits:

- Denial of sexual, emotional, cultural, ethnic or racial complexity;

- Covering up uncomfortable or unpopular political reality;

- Political strategy against dissidents or free thinkers;

- Hegemonic strategy used to publicly pillory the 'outgroup';

- Fascist strategy to curtail down civil liberties for social scapegoats.

THE DESTRUCTION OF THE NATURAL ORDER

What is Moralism?

Thomas Moore writes in his book *Care of the Soul (1994)* on the subject of moralism:

> Moralism is one of the most effective shields against the soul, protecting us from its intricacy. (…) I would go even further. As we get to know the soul and fearlessly consider its oddities and the many different ways it shows itself among individuals, we may develop a taste for the perverse. We may come to appreciate its quirks and deviances. Indeed, we may eventually come to realize that individuality is born in the eccentricities and unexpected shadow tendencies of the soul, more so than in normality and conformity. (Id., p. 17)

> Care of the soul is interested in the not-so-normal, the way that soul makes itself felt most clearly in the unusual expressions of a life, even and maybe especially in the problematic ones. (…) Sometimes deviation from the usual is a special revelation of truth. In alchemy this was referred to as the opus contra naturam, an effect contrary to nature. We might see the same kind of artful unnatural expression within our own lives. When normality explodes or breaks out into craziness or shadow, we might look closely, before running for cover and before attempting to restore familiar order, at the potential meaningfulness of the event. If we are going to be curious about the soul, we may need to explore its deviations, its perverse tendency to contradict expectations. And as a corollary, we might be suspicious of normality. A facade or normality can hide a wealth of deviance, and besides, it is fairly easy to recognize soullessness in the standardizing of experience. (Id., 18)

Moralism in Education

Traditionally, in patriarchal culture and society, education was moralistic. But even in our days of feminism and open criticism of patriarchal tradition, moralistic education has survived. It often takes hidden forms.

Often concepts that are outspoken intellectual are forms of ideological pressure the child will be submitted to in the name of their own best, and for ultimate compliance with social and political dogma.

The suppression of the child's emotions has many names and takes many subtle forms. It is manifested also in the intellectual dressage of the child. Who thinks only does not feel much, or much less.

Such kind of water-head education as practiced by *Montessori* schools may produce good surgeons or computer programmers.

—See: Maria Montessori, The Absorbent Mind (1995).

However, happy and harmonious human beings who think ecologically and can help healing the earth do certainly not come out of such educational institutions. Many of them will be active to bring

The Destruction of the Natural Order

about further destruction and misery to this tortured and moralism-enslaved humanity.

Repression and Perversion

Repression and Denial

Repression is a term coined by Sigmund Freud that describes a function of the psyche in the case consciousness meets a desire that is strongly prohibited by the inner controller.

> —Eric Berne recognized three essential inner selves: Inner Child, Inner Parent and Inner Adult. In my own research and work with the inner dialogue during a two-years Erickson hypnotherapy, I encountered the presence of additional entities such as the Inner Controller or Inner Critic as the instance in the psyche that represents the societal, cultural and moralistic values that we have internalized through education and conditioning. If the Inner Controller is hypertrophied and thus dominating the psyche, the result is that we are unable to realize our love desires.

What happens in this case is that the psyche will repress the desire into the unconscious in order to uphold the functioning of the ego which would otherwise be disturbed in maintaining the integrity of the psyche.

Natural Order

—The function of the ego is not to dominate any of our inner entities, but to orchestrate them, to direct them in a team-like cooperation, such as for example the conductor of an orchestra leads more than one hundred musicians to play in sync in order to reproduce a musical score with accurate precision and harmonious sound. This is the function of the healthy ego within the multidimensional psyche. Needless to add that with most people the ego and the inner controller are hypertrophied and dominate if not suppress all other inner entities which is the explanation why such a high percentage of the world's population is completely uncreative, dull and imitative in their behavior, and why they use only about eight percent of their emotional intelligence potential.

On the starting point of repression is denial. All repression starts with a denial, the denial of one of our inner entities or energies, as they are part of our *inner selves*. Denial is mainly the result of moralism, the quite arbitrary and culturally conditioned split between good and bad emotions, good and bad feelings, good and bad desires, good and bad behavior.

Repression and Regression

It has to be seen that *regression*, while a familiar term in psychology, is entangled with *repression*. *Every form of repression results in regression.* When I repress my sexual desire for copulating with adult women, for whatever reason, I will regress in my sexual maturity, and the result is that I will turn the wheel of sexual evolution backwards, to enter the realm of the paraphilias. In most cases, I will not be consciously aware of the fact that a repression takes place, or that I deny my desire. I will not be aware that I block the *emotional flow*, and that I am in a fear-condition.

Regression in psychology is not usually linked to repression. While repression always leads to regression, regression can also occur independently of repression, and then we talk about an entirely different set of phenomena. Regression in psychology and in natural healing typically is the fact of leading the patient back to the original wounding as from a point of effectiveness and natural psychosomatic dynamics, without the *encountering of the original wounding* in the dream, hypnotic or trance state, a real and full healing can generally not be achieved.

Repression and Retrogradation

Repression results in the bioenergy that feeds this desire to retrograde and change polarity. I have adopted the term *retrogradation* from astronomy and astrology, and was inspired insofar by the humanistic astrologer Dane Rudhyar as it is really a lucid metaphor: retrogradation means that the energy of the planet becomes introvert for the time of the retrogradation.

And the similarity to psychological retrogradation is striking! When a child's *primary bioenergetic vitality* is impaired by moralistic education, the child looses spontaneity and becomes shy and introvert, thereby dramatically reducing communication strings with the outside world, and at the same time begins to communicate more strongly on the inside level.

This leads to timidity and can result in an impaired communication ability for life, such as stuttering or extensive sweating when being around people. In the extreme case, under conditions that amplify the original retrogradation of the natural drive for full emonic satisfaction, sadism begins to develop and can become an obsession that dominates the whole sexual life of the person. Another consequence that repression brings about, is regression, which means a devolution of consciousness, a backward development in terms of psychosexual maturity—a

THE DESTRUCTION OF THE NATURAL ORDER

state that is undesirable socially, because it involves a reduction of consciousness.

Repression and Projection

Projection is a psychic automatism that is a by-product of repression. When an emotion or desire gets repressed, projection sets in and what is blinded out from wake consciousness is projected upon others – who then get the blame for what is originally a part of the person's own life. This is the true meaning of Matthew 7:1-5 which says:

> Matthew 7:1-5
>
> Judge not, that you be not judged. 2 For with the judgment you pronounce you will be judged, and with the measure you use it will be measured to you. 3 Why do you see the speck that is in your brother's eye, but do not notice the log that is in your own eye? 4 Or how can you say to your brother, 'Let me take the speck out of your eye,' when there is the log in your own eye? 5 You hypocrite, first take the log out of your own eye, and then you will see clearly to take the speck out of your brother's eye.

Cultural Perversion

Perversion, in a general non-moralistic sense, is to put nature upside-down, to replace natural healthy organismic processes by artificial unhealthy mechanical processes. In a metaphorical sense, perversion is the image of the dethroned, ravished and reversed goddess, or the reversed *Lunar Bull* as her traditional consort.

The quintessential example of a perversion is the *repression* of natural desires because they are judged unwanted under a certain ideology or contrary to well-defined forms of conduct. What then happens is namely that the bioenergetic continuum and equilibrium that is part of all natural desires is disturbed or disrupted, and the result is a reversal of bioenergetic polarity that brings about a retrogradation of the original impulse.

This retrogradation is the actual perversion of the impulse. The following quote from Emerson's essay *Compensation* says it very clearly:

> RALPH WALDO EMERSON
> The history of persecution is a history of endeavors to cheat nature, to make water run up hill, to twist a rope of sand. It makes no difference whether the actors be many or one, a tyrant or a mob. A mob is a society of bodies voluntarily bereaving themselves of reason and traversing its work. The mob is man voluntarily descending to the nature of the beast. Its fit hour of activity is night. Its actions are insane like its whole

THE DESTRUCTION OF THE NATURAL ORDER

constitution. It persecutes a principle; it would whip a right; it would tar and feather justice, by inflicting fire and outrage upon the houses and persons of those who have these.

—Ralph Waldo Emerson, The Essays (1987), p. 69. (Compensation)

Perversion appears to be produced by fear. And it is equally true that psychological fear is perversion, an upside down of the *élan vital*, a retrogradation of the love energies, an obstruction of the life force.

The most important thing to know about perverse desires is that they come about *through the repression of original desires;* thus, the perverse desire kind of *replaces* the original desire and *compensates* for its lack. In other words, the perverse desire has two functions, a *replacement function* and a *compensation function*.

Perversion, we could attempt to define, then, is a strongly distorted form of sexual love, a sexual desire that is mutilated in a way to result in its very contrary. Instead of love and life, what comes out in perversion is hate and death. In the Freudian terminology, we would say that perversity is not a form of libido but a variant of the death instinct.

NATURAL ORDER

LEGISLATIVE PERVERSION

Sex laws, the laws that persecute lovers for certain forms of love, are truly a *legislative perversion*. One may argue that rape is a perversion from natural consenting love, and I think much speaks for that view, but it does not help us to understand sexual paraphilias in general, and pedophilia, in particular.

Rape occurs equally in the marriage bed and supposedly much more often as it happens in pedophile relations. Thus, even if one may consider rape as an act so hostile and so far removed from normal love that it has in itself to be considered a perversion, this is by no means a criteria to be applied only for adult-child erotic relations.

Certainly, some pedophiles rape, but also some heterosexuals rape, and some homosexuals rape. Rape has reasons that are not related to a person's sexual orientation. It has primarily power reasons.

—Nicholas A. Groth, Men Who Rape (1980).

Rape is a form of compensation for feeling powerless that uses sexuality as a weapon against another rather than a loving exchange as it is in normal sexuality. I think when we seriously consider these implications, we can even learn a lesson about human love, namely by looking at its very contrary.

Krishnamurti once said that we cannot define love while we can define and look at all what is *not* love. Looking this way at perversion, we can see that in all perverse behavior there is a residual form of cynical love that, if it was not so tragic, could be considered as clownish in some way, or scurrilous. When we look deeper, we encounter fear, much fear. We can then see that it is fear, and nothing but fear that originally distorted love into perversion. In love there is no fear; love is carefree, love is abundant and it is giving.

Perversion is paranoid, it is avaricious and takes only, unable to give, utterly narcissistic. Love is sharing, and shared pleasure, while perversity is egotistic and lonely enjoyment at the cost of another, even at the cost of their life. Thus, while in love there is always natural care, perverted love typically is little or not caring about the love mate.

What is the fear, then, or what are the fears that distort natural love into various forms of perversion, and violence? A generalized answer is difficult to give. There are complex reasons, individual and collective.

NATURAL ORDER

Since childhood I doubt that ours is a natural and loving culture. I always found our culture unnatural and perverse in its very roots. To prohibit the child to live their natural emotions and sexuality is perverse. There is no argument to be brought up against that. This millenary practice in itself is a flagrant violation of nature in its most tender origins.

Another example for a true legislative perversion is the fact that most governments have declared certain plants illegal, calling them *drugs* – persecuting their use and even their mere possession. As Terence McKenna rightly pointed out, *it borders mental derangement to declare certain plants illegal.*

And on the same line of thought, I call perversion the fact that most governments, based upon so-called child protection laws, can now declare writings, publications, books, as illegal, thereby reinstituting the arrogant practice of the Church that used the burning of books as one of many means to suppress truth.

The Church called any derivation from the official dogma *heresy*; well, today it's called *child pornography*. It's exactly the same thing, under a different name.

THE DESTRUCTION OF THE NATURAL ORDER

Another example. When pedoemotive desires are strongly repressed, as it is the case in most Western societies, and so much the more when such behavior is not socially coded, there is a danger that the original tender and caring erotic desire for love with children turns upside down; that means that the vital energies contained in the desire retrograde and turn into a violent, disruptive, abrasive, cynical and dominating drive out to achieve momentary or prolonged sexual satisfaction with a child through abduction, rape, torture and abasement of the child. The perversion in this case is not, as moralists tend to argue, the original pedophile desire, but its very *repression* and modern culture's persistent denial to code these desires socially. It's through repression and denial that these originally peaceful desires turn violent.

As a result of this insight, I claim that postmodern consumer culture, through the mythic construct of child protection and the resulting fascist denial of democratic exchanges and erotic love between adults and children, and through a fervent denial to code nonviolent childlove *actually endangers children and exposes them to constantly raising threats of sexual child abuse, child abduction and child murder.*

This is the very contrary of caring for the safety of our children! Statistics show with striking evidence that in developing countries children are safer from abduction and sexual abuse and than in any of our highly *child-protective* Western consumer societies.

One of many facts that corroborate my hypothesis is that the country where the polemics on *child protection* is highest, where the taboo on adult-child sexual relations is strongest and where sex laws are the most Draconian on earth is the United States of America. And it is the United States that shows among all nations on earth by far the highest statistics of child abduction, sexual child abuse and child murder. This is really the *culmination point of social perversion* because it's this country's government that goes around to allegedly better the world, and to play the eternal judge on human behavior and morality worldwide.

THE DESTRUCTION OF THE NATURAL ORDER

Religious Perversion

Calvinism

Calvinism was an atrociously extremist perversion of the Christian dogma in its Protestant vintage. It was brought up by the French Swiss *Jean Calvin (1509-1564)*, a lawyer and fanatic protestant reformer. Calvinism is best known by the tortures it has inflicted upon children and even infants, to withhold them from masturbating, thus attaching their tiny hands to the bed's wooden frame, which caused in some cases long-term paralysis and even death of the infant.

The horror of these tortures is described in many studies, and was carefully reviewed and courageously claimed as abysmal child abuse by the American psychohistorian Lloyd DeMause and the Swiss psychoanalyst Alice Miller.

—Alice Miller, Thou Shalt Not Be Aware (1998) and For Your Own Good (1983).

Puritanism

Puritanism began in the 16th century during the reign of *Queen Elizabeth I* as a movement for religious reform. The early Puritans felt that the Elizabethan ecclesiastical establishment was too

NATURAL ORDER

political, too compromising, and too Catholic in its liturgy, vestments, and episcopal hierarchy.

> —See, for example, William Haller, The Rise of Puritanism or The way to the New Jerusalem as set forth in pulpit and press from Thomas Cartwright to John Lilburne and John Milton, 1570-1643, London, 1938; Christopher Durston and Jacqueline Eales (Eds.), The Culture of English Puritanism, 1560-1700, London, 1996; John Spurr, English Puritanism, 1603–1689, London, 1998; The New England Primer, Improved for the More Easy Attaining the True Reading of English: To Which Is Added the Assembly of Divines, and Mr. Cotton's Catechism, 1991; E. S. Morgan, Visible Saints (1965); J. E. C. Hill, Society and Puritanism in Pre-Revolutionary England, 2d ed. 1967); H. C. Porter, Puritanism in Tudor England, 1970; C. L. Cohen, God's Caress: The Psychology of Puritan Religious Experience, 1986; C. E. Hambrick-Stowe, The Practice of Piety, 1986; S. Foster, The Long Argument: English Puritanism and the Shaping of New England Culture, 1570–1700, 1991.

Calvinist in theology, they stressed *predestination* and demanded scriptural warrant for all details of public worship. They believed that the scriptures did not sanction the setting up of bishops and churches by the state. The aim of the early Puritans such as Thomas Cartwright was to purify the Church (hence their name), not to separate from it.

After the 17th century, the Puritans as a political entity largely disappeared, but Puritan attitudes and ethics continued to exert an influence on American society.

They made a virtue of qualities that made for economic success such as self-reliance, frugality, industry, and energy, and through them influenced modern social and economic life. Their concern for education was important in the development of the United States, and the idea of congregational democratic church government was carried into the political life of the state as a source of modern democracy.

The Inquisition

The *Inquisition* was organized murder perpetrated by the Church in an unmatched holocaust in which for the most part young women and girl children were persecuted as *heretics*, tortured in unspeakable ways and put to death by burning. This plague of religious perversion and violence lasted for several centuries in Medieval Europe and was never really labeled by any modern human rights movement as what it truly was: the first organized global genocide in human history.

NATURAL ORDER

Love and Split-Love

Love or Loves? is the pertinent question I am asking since my childhood. I believe the natural order knows love, the perverted order knows loves. It's as simple as that.

Contrary to common understanding, I emphasize a *holistic understanding of love* that I express under the header 'love is unity.' That means that love is always erotic, and erotically intelligent. To my knowledge, this understanding of love is novelty and has not been introduced yet as a scientific or philosophical concept. Yet all great poetry implicitly expresses this truth. Succinctly speaking, this means that I am against the splitting off of love into so-called *erós* and *agapé*, on one hand, and the further splits of the unity of love into neat concepts.

I believe that the natural order knows this unity of love, which is one element why it was a paradigm that fostered peaceful human togetherness, not dominance, violence and war.

Usually, our encyclopedias denote conceptual notions of love and enumerate them as:

THE DESTRUCTION OF THE NATURAL ORDER

- parental love;

- family love;

- motherly/fatherly love;

- love of children for their parents;

- siblings love;

- love for the ancestors;

- love for one's home country or patriotic love;

- love for tradition;

- passionate love;

- love for one's husband or wife;

- and so on and so forth.

Where does this reductionist concept of love lead to? In my analysis of this question, that I asked for the first time in high school, in my philosophy class, the answer why this happens is the fragmentation of modern man and the rationalization and intellectualization of love.

—See also Michel Odent, The Scientification of Love (1999).

NATURAL ORDER

What originally is a matter of the heart became a concern for the brain, and instead of letting go for love to come as a spontaneous, novelty kind of thing, people in modern societies tend to think about love and wish to be loved instead of simply loving, and without asking anything in return.

The intellectualization of love, while it's a rather modern phenomenon, is the result of splitting love in permitted and forbidden love, which is based on the upsurge of *compulsive morality* throughout patriarchy. Already long before industrialization, Christian life denial has done its part in the destruction of natural love and its more or less total perversion into the love-and-sex dichotomy that today is part of mainstream sexology.

The very split of love into love, on one hand, and sex, on the other, is perverse and anti-nature. The reason why this schizoid split was created in the psyche and behavior structure of Western people is moralism and fear, and here, in particular, fear of incest.

When a father says that he loves his little daughter, people want to make sure that this father means he loves his daughter in compliance with the incest

THE DESTRUCTION OF THE NATURAL ORDER

taboo, and not as a full sexual mate. However, a unifying concept of love says that love always contains the potentiality of sexual attraction. My concept of *emonic attraction* is indeed such a *unifying concept of love* that contains no moralistic element, thus giving nature full credit.

There is no need to pervert our language simply because we are afraid that parents and children may not only experience love but also sexual attraction for each other. The chances that people act out on these attractions are, according to statistics, after all not very high, and yet because of various reasons, these ideas have become group fantasies and mass obsessions for modern man and are for this reason, and for this reason only, in our daily press. The correct way to use language, and to use the word love, is to imply in it all its potential meanings and connotations, instead of cutting them out by splitting off love into loves, thereby destroying the unity of love. For there is no way out of this split-concept than the *antithesis of perversion*. We need to accept reality instead of fighting reality. Love is like the sun. It is impartial, and the force of its irradiation does not depend on those who are bathed in the heat waves. With language it should be alike.

When I say I love children I don't bother if you think that I also love them erotically. What you think is your business, not mine. But we have to keep language pure so that our *code* doesn't get messed up. I would rather say that the burden of proof in this case is upon you to demonstrate that you do *not* love children erotically—and why! (When love is unity, we by default love children also erotically!)

When you deny children their capacity to be erotic for adults, you deny their vital energy, and thereby you deny them to truly be alive and live! This is simple truth and not a propagandistic statement, if you can accept that in your sociocultural alienation and conditioning is not my problem, but yours!

The burden of proof is upon death to disprove that life is *unity of life and death*.

The Disintegration of Sexual Paraphilias

In psychology and sexology, *paraphilia* is a term that describes sexual arousal in response to sexual objects or situations *which may interfere with the capacity for reciprocal affectionate sexual activity.* Paraphilia may also be used to imply non-mainstream sexual practices without necessarily implying any dysfunction or moral deviance.

Gerontophilia

Gerontophilia literally means love for the elderly, or, if a child is concerned, for adult intimate partners. Gerontophilic desires are mirroring pedophilic desires and manifest and express love and sexual attraction for older people than oneself.

Gerontophilic emotions can be present in adults or children. For example, workers in institutions for the elderly may at times experience attraction to one or the other elder patients. Children experience residual attraction toward adults, probably not for all adults they consider as caretakers or close friends, but for some of them. However, it has to be seen that this potential attraction is amplified and crystallized through the Oedipal trap.

Generally speaking, much is unknown in this area as child psychology and sexology, under the spell of the rigid mainstream paradigm, have till now not been able to reveal the cognitive framework that comprehensively explain our love choices in general, and the love choices of children, in particular. It appears that gerontophilia is a *conditioned response* in children reared in societies that do not allow children to live their sexuality, early in life, with peers; thus what happens is that children's emotional and sexual energies are channeled toward their parents.

Freud's assumption the *Oedipus Complex* was universal is in the meantime refuted by anthropological research. In tribal societies where children enjoy uninhibited peer sexuality, they are not or very little gerontophilic, and in fact in these societies, for example the Trobriand islands, pedophilia and gerontophilia have been found equally non-existent by the field researchers Bronislaw Malinowski and Margaret Mead as early as in the 1920s.

THE DESTRUCTION OF THE NATURAL ORDER

Gerontophilic feelings and tendencies are in psychoanalysis discussed under the header of 'Oedipal' desires. Regarding other adults than the parents, caretakers, educators, friends or even strangers, psychoanalysis was and is reluctant to admit children's gerontophilic emotions. Among the few professionals who were outspoken about the desires, was the French psychoanalyst and child therapist Françoise Dolto (1908-1988).

Pedophilia or Childlove

My approach to *pedophilia* is markedly different from mainstream psychology, sexology and the mouthpiece media of the international child protection industry in that I acknowledge the hitherto suspiciously overlooked existence of *pedoemotions* as a universal form of adult-child attraction that is biologically programmed for the continuation of the human race.

Pedoemotions are a *natural manifestation of pedophilic attraction*, without however being per se sexual – as the word says. They are namely first of all *emotional*, as the primacy in attraction is emotional, not sexual. And I say this in full consciousness that it's against the mechanistic setup of sexology research.

Pedophilia then, is to be considered as a condensation and sexualization of pedoemotions. Why pedoemotions can crystallize into an exclusive emotional attraction for children that becomes sexualized is not yet known, and I keep away from daring speculations. Many hypotheses have been aired that show more of the individual bias of the researcher than about anything else. The mainstream ones stress that early abuse with the reverse effect of *abused becoming abuser* is to be found in a high number of cases. But of course, forensic research always only investigates cases that go to court, and thus where things have gone wrong and relationships went south, in one way or the other.

Lesser known theories come up with genetic and karmic factors, or consider the possibility that pedophilia may be a hidden form of incest (Lloyd DeMause). The latter hypothesis seems to contradict the fact that pedophilia is by definition a relationship between an adult and a child not affiliated with each other, as in the contrary case we speak about incest in direct or indirect line. To mix up and mess up this clear distinction means to mix up and mess up dream and reality. One may well fantasize about incestuous relations, *but that does not make one a pedophile!* In fact, much research on heterosexual men and women has been undertaken that found a considerable amount of incestuous fantasies during masturbation, and yet these men and women were and remained *heterosexual* in their overall sexual behavior, and did not suddenly turn into pedophiles just because of their fantasies. The fact of men or women being exclusively sexually attracted to pre-pubescent children, refusing to have sexual conduct with an adult is something very seldom to be found in tribal cultures.

In my research on native cultures and shamanism, I found references for such behavior, but in every case the person that the author was referring to was considered to be the village idiot, a marginal freak that most people did not take serious and that often is allowed to have his way with those children who are equally marginal, typically mentally retarded or handicapped children. And last not least, we should consider that human beings are not automatons and constantly make *love choices*, which means that they vote for options.

What I am saying is that *there is a certain amount of conscious choice* involved in the fact that a person becomes as it were a pedophile, for years or even decades of their lives. There may be a love choice for children, or for boys only, or girls only, as a result of emotional frustration, a broken marriage or many broken or accidented heterosexual or homosexual relations with adult partners. This has never been considered in research and this is logically so because sexology is completely mechanistic and considers human sexual attraction more or less as a *set of automatisms* that result from early sexual conditioning.

I believe there is choice involved in creating a *pedophile identity*. While this identity may be a fake identity as it is based solely on sexual attraction, and not upon soul values, it's an identity nonetheless. Why men and women choose this identity is hardly known, and it's difficult to comprehend because suffering surely is involved in this choice as current society's virulently rejects this kind of emosexual attraction.

THE DESTRUCTION OF THE NATURAL ORDER

I believe we will gain much insight in the human nature, and in love, once we begin to dare revisiting pedophilia research with this special perspective and focus in mind. It may also lead to a greater understanding of men and women who have done such love choices, despite the often painful social situation that this involves.

I am saying that *sexology knows nothing of the etiology of pedophilia*; this is so because the truth can only be found through research on the human energy field and not on sexual behavior, as I suggest is with *Emonics*. This is so because sexual attraction follows emotional addiction, and not the other way around.

So the answer if the hen or the egg was first can be answered clearly in this case: first comes emotional attraction, and then, if ever, such attraction may become sexualized for reasons that we still ignore.

Boylove or Pederasty

Boylove is the eroticized friendship and love of adult men with pubescent and with lesser likelihood pre-pubescent boys that is idealized to a certain extent, and where sexual interaction is not sought after as a primary means to engage the relation, but as an add-on in a more encompassing relationship that is guided by a pedagogical interest of the boylover for the educational welfare of his love boy.

Boylove has been reported from *all ancient cultures around the world,* since the beginnings of written history. While it has been repressed with equal universality by all major religions, it survives in subtle forms and is today better organized than the girllove movement.

One substantial point in the agenda of both boylove organizations and homosexuals is to explain to the public that boylovers are not homosexuals, and that homosexuals typically have no interest in boys, as the general public generally tends to confuse homosexuality and boylove.

Another important point in the public agenda of boylove organizations for example in the United States, Holland or Germany is that they stress that they are *not rapists and sex offenders,* but that their love is based upon respect for the boy and consent from the side of the boy.

In a 1982 study at the Faculty of Psychology of the University of Utrecht in Holland, conducted by

THE DESTRUCTION OF THE NATURAL ORDER

Professor Dr. Theo Sandfort, a sample of twenty-five man-boy couples was taken and investigated, through interviews of both the boys and their lovers, and the results showed a surprising incidence of non-conflictual parameters. In fact, in one third of the cases, the parents of the boy were informed about the relation and allowed it to continue.

In the other cases, a slight but not pathological amount of stress has been noted that was caused by the fact that the relation had to be kept secret both in front of the boy's parents and society at large. But the most surprising outcome of the study was that the sexual aspect of the relation was not causing any significant amount of stress, let alone harm, and in all relations the boys fully consented to sex, while the general tenor was that sex is not the most important part of the relation, but rather values like emotional closeness, security, trust and shared activities. Regarding sexual activities, it was clearly established that penetration had occurred only in a minority of the cases and that in most cases the sexual behavior was of a masturbatory kind, including fondling and kissing.

NATURAL ORDER

ON THE EXISTENCE OF NEPIOPHILIA

The difference between pedophilia and nepiophilia is that in nepiophilia the love object is considerably younger, typically a small child, baby or toddler. *Paidós*, in old Greek, means child or boy, but a child that is no more a baby. *Nepiós*, typically and by distinction from *paidós*, is a baby or toddler.

There has been a phase in my life, more than twenty years ago and while an unhappy childless marriage was coming to its end, that my sexual orientation turned very strongly toward young children of both sexes. It was not an exclusive attraction, but the joy that being around small children procured me at that time was so unique, so unusual, so great and so fulfilling that I was opting for a new career as an early child educator.

The subtle eroticism in my relations with small children and toddlers was distinct from what I formerly lived as so-called normal sexuality with my wife – very different. First, it was certainly not sexuality in the sense of the word, but a *sensual interaction* that involved subtle feelings to be qualified as erotic, but not as sexual, accompanied by strong excitement, or enthusiasm, and sometimes temporary sexual arousal.

THE DESTRUCTION OF THE NATURAL ORDER

It was certainly not something even remotely similar to going to bed with a person and having an orgasm, but something ongoing and so subtle that nobody being present in the same room would ever even notice it. It was also an excitement that is pregenital, so to say, in that in most cases it is not marked by a phallic erection. This experience actually led me to later discover the existence of *pedoemotions*.

Often times, during that phase of my life, I was wondering if not women, mothers, female babysitters and teachers do in fact have similar feelings, so much the more as mine did not result in an erection, but were in character more of what I had read in female reports about sexual excitement, something namely that is felt *in your whole body at once, much like an electric current*. For example, when I oiled a baby boy's body after his bath—he was at that time around twenty months old—and he smiled at me, I also felt that current in my fingertips and my hand palms. I think the boy has been *energized* through that tender massage because sometimes, he jumped up from the baby table and put his arms up like a little gorilla, full of strength and joy.

NATURAL ORDER

In clinical studies and statistics about pedophilia or sexology in general, it is often stated that nepiophilia was a very marginal phenomenon in human sexuality. I doubt these findings on the basis of the evidence, the observations and personal statements and views that I have accumulated throughout the years of my research.

Today, I tend to think that nepiophilia has the same importance as general childlove within the total range of pedoemotions that the human being is invested with by nature. Apart from that more general view, I can make out a *certain dynamics in our sexual drive that is functional* and dependent upon the tasks we wish to fulfill in life.

For example, a new parent is naturally more invested with nepiophile emotions than with pedophile ones, simply because nature wants them to take care of their small child. Once that boy or girl is a budding adolescent, the father may perhaps think back of that toddler with a strange feeling of 'how could I possibly be so crazy about that little child at that time?' Because now our good father is crazy about his budding school girl! And that is good so and should be so. It is the very reason why we

THE DESTRUCTION OF THE NATURAL ORDER

procreate, because the life function is a pleasure function.

Childlove in Literature

Johann Wolfgang von Goethe, while he had himself pederastic inclinations, that he openly revealed in many of his letters, found it probably too daring and offensive to present the original *Faust* tale to the German bourgeoisie of his time. Goethe's *Faust* is thus a variation, and to a certain extent a falsification of the original *Dr. Faustus* plot that was quite different and had nothing to do with a man-woman relationship. In the original tale, Dr. Faustus was namely a school teacher and pederast who was imprisoned for sexual relations with some of the children in his class.

Goethe's Faust however counts a new tale involving a young girl named *Gretchen* who was dishonored by Faust. It's interesting that Goethe maintained in his version the element of a sexual offense, and this detail is indeed significant because in these times any form of non-marital sexuality was considered by the Church as being instigated by the devil or devilish instincts.

NATURAL ORDER

Death in Venice by Thomas Mann is a clear account of pederastic feelings of a German academic and poet called *Aschenbach* toward a beautiful 14-year old Polish boy with the name of *Tadzio* he met in a resort hotel on the Lido in Venice. The base idea of the story is most probably autobiographic. The tale ends with the outbreak of cholera and Aschenbach's sudden death despite a highly alarming premonitory dream. In fact, he had already packed and went to the railway station in time, but returned because sudden strong desire instigated in him the wish to seduce the boy sexually. His desire ended in a fatal way, and that bad end of the otherwise enthralling and erotically appealing story tells more about its author than about anything else. In fact, in many of his writings, Mann subtly reveals that he had strong feelings for boys that he however never admitted to realize during his lifetime, to appear as the perfect citizen. This attitude was criticized after his death as being hypocrite and false, as it occurred with an artist and poet.

Paul Goodman (1911–1972) was an American poet, writer, and public intellectual. Goodman is now remembered as a notable political activist in the 1960s and early 70s. Politically he described himself as an anarchist, sexually as bisexual, and professionally as a man of letters. Less widely known is his role as a co-founder of *Gestalt Therapy*. In this regard, he was strongly influenced by Otto Rank's *here-and-now* approach to psychotherapy, fundamental to Gestalt therapy, as well as Rank's post-Freudian book *Art and Artist (1932)*.

The freedom with which he revealed, in print and in public, his homosexual life and loves proved to be one of the many important cultural springboards for the emerging *gay liberation movement* of the early 1970s. However, his own views ran counter to the modern construction of homosexuality. It was his opinion that it was pathological not to be able to make love to someone of the opposite sex, but that it was equally pathological not to be able to experience homosexual pleasure. Likewise, it was his view that sexual relationships between men and boys were natural, normal and healthy, and that they could lay the foundation for continuing friendship even after the sexual response is outgrown.

André Gide (1869-1951) was a French author and winner of the Nobel Prize in literature in 1947. Gide's work can be seen as an investigation of freedom and empowerment in the face of moralistic and puritan constraints, and gravitates around his continuous effort to achieve intellectual honesty. His self-exploratory texts reflect his search of how to be fully oneself, to the point of owning one's sexual nature, without at the same time betraying one's values. His political activity is informed by the same ethos, as suggested by his repudiation of communism after his 1936 voyage to the USSR. Known for his fiction as well as his autobiographical works, Gide exposes to public view the conflict and eventual reconciliation between the two sides of his personality, split apart by a straightlaced education and a narrow social moralism—as he perceives himself: the austere and refined Protestant, and the divinely inspired – and no longer blushing – pederast.

Yukio Mishima was the public name of *Kimitake Hiraoka (1925-1970)*, a Japanese author and playwright, famous for both his highly notable nihilistic post-war writings and the circumstances of his ritual suicide.

Mishima began his first novel in 1946 followed by *Confessions of a Mask*, an autobiographical work about a young latent homosexual who must hide behind a mask in order to fit into society. The novel was extremely successful and made Mishima a celebrity at the age of twenty-four. Mishima was a disciplined and versatile writer. He wrote not only novels, popular serial novellas, short stories, and literary essays, but also highly-acclaimed plays for the Kabuki theater and modern versions of traditional Noh drama.

Although he visited gay bars in Japan, Mishima reportedly remained an observer, and had affairs with men only when he traveled abroad. After briefly considering an alliance with Michiko Shoda—she later became the wife of Emperor Akihito—he married Yoko Sugiyama in 1958. Over the next three years, the couple had a daughter and a son.

These examples are by no means exhaustive. Also note that in older literature, and in non-Western literature, the present-day clear distinction between pederasty and homosexuality was not that clear-cut and often blurred. Especially with Yukio Mishima this is the case, as many of a 'young man' in the writings of this and many other writers would today be called a boy. So the label *homosexual* attributed to many authors has to be questioned. Things are not that clear-cut in our feelings, emotions and sexual attractions, and life's intrinsic erotic complexity is dangerously reduced by such kind of labeling and self-labeling. In Goethe's case things are much more evident in this respect as it's clear from his letters and otherwise documented that he for example had a long-term love relation with a 6-year old boy who was the son of Baroness von Stein, one of Goethe's lifelong friends and admirers.

Parent-Child CoDependence

I discuss parent-child codependence synonymously under the headers of cofusion, secondary fusion, pseudo-fusion or codependence.

Codependence is a dependency problem that manifests in the parent-child relation typically for the first time after the critical mother-infant symbiosis, and thus as a general rule after the 18th month of the baby.

What is generally very little known is the fact that even before the completion of the 18th month of the infant, mother and child are engaging in subtle communication *about limits* that typically reveals to what extent the mother is able and willing to grant the infant autonomy, or not. This early dialogue, that is most of the time nonverbal, has been found to deeply condition people for their later relational behavior patterns.

This is even more relevant in the mother-son relation than it is in the father-daughter relation because the *matrix-provider* has more power of the child, be it boy or girl, than the *sperm-giver*. This evaluation of the primal scene has been found both by Freudian Transactional Analysis (TA), and it is not as such a matter of cultural conditioning, or compliance to either matriarchy or patriarchy.

Co-Dependence is a major building block in the political and social entanglement scheme of what I call *Oedipal Culture*. Causative factors that have been revealed in my own and other research are:

- Mother did not really want the child;
- Mother is professionally over-engaged, lacking time for the infant;
- Lack of healthy physical interaction between parents and child;

- Overly strong career focus of parents, leaving child to babysitters;

- Insufficient eye contact in the mother-infant relation;

- Insufficient or no breast feeding;

- Insufficient tactile stimulation of the baby (tactile deprivation);

- Shame-based identity of the mother and resulting rejection behavior:

    - when baby shows erotic behavior, and mother turns away regard;
    - when baby touches his or her genitals, and mother takes their hands off;
    - when baby seeks closeness with mother, she puts baby to sleep;
    - when mother holds baby away from her body, to avoid touch;
    - when mother constantly has 'no time' for intimate time with baby;
    - etc.

- father left family during pregnancy, after birth or not long thereafter;

THE DESTRUCTION OF THE NATURAL ORDER

- father, while still part of the family, is as good as never present;

- father refuses to take over any role in childcare;

- father is abusive toward mother and/or the child, etc.

In other words, *codependence is a compensation reaction of entangled organisms* that tries to heal a split that was caused by a lack of early intimacy.

The entanglement paradoxically comes about *through a lack of physical closeness*, and of communication, and through a general tactile deprivation of the child, also through non-physical factors such as parents' thoughts being constantly focused on money and status or children generally relegated to receiving affection from secondary caretakers, babysitters, house teachers, and the like.

The entanglement specifically comes about through the fact of lacking autonomy of the child, and of lacking exposure to experiences and a social life outside of the family.

Details have been shown with abundant evidence by the long-term research of James W. Prescott,

Ashley Montagu, Michel Odent, Frederick Leboyer, Alexander Lowen and others. The problem of co-dependence is for obvious reasons much more stringent in the individualistic and separative white Western cultures than in highly sociable open societies such as African, South American or Asian cultures. Yet in these cultures today we face the problem in the middle and upper classes as well because they have adopted Western values and a lifestyle that clones most of the alienated Western behavior models, thereby shunning their own perennial wisdom that their elders still are knowledgeable about.

The Popular Confusion

There are many myths that distort and tear down naturally erotic but nonsexual relations between parents and children; these distorted popular views actually foster and purport codependence instead of helping in any way to avoid it. For example, and contrary to popular belief, the unhealthy co-dependence between parent and child is not created through too much of physical interaction and shared affection and tenderness, but in the contrary

THE DESTRUCTION OF THE NATURAL ORDER

through touch hostility and prudishness. For example, it has often been believed that a boy will develop a codependent relationship with his mother when he is 'too close' to her, or when he sleeps with his mother in the same bed. This is simply not true. The causes of mother-son codependence are often depicted in an overly simplified or even distorted manner.

To begin with, it is not through abundant shared pleasure, affection, tenderness and touch that codependence comes about. It's not through mother and son, or father and daughter, sleeping together, taking baths together, sharing nudity, and it's not through their sharing a naturally sensual and erotic attraction for each other.

In the contrary, if these elements were causative factors in the etiology of codependence, any abundantly sensual mothering or fathering would lead to entrapping children in pseudo-incestuous relations. But this is not the case. If a mother is fully erotically present for her boy, without being incestuous, and embraces him sensually while giving him at the same time the necessary amount of autonomy according to his age and abilities, the boy will easily liquidate his *Oedipus Complex* and develop his fully functional heterosexuality; he will then project his libido upon peer girls of his age, or approximately of his age.

The same is true in the father-daughter relation with regard to the girl-child's mastering the Electra Complex and projecting her sexual feelings upon peers boys.

There are many false signals in today's popular culture and vulgarized psychological publications. These false signals lead to parents' becoming more and more insecure as to the role physical affection plays in parenting. This makes that parents are more or less constantly bombarded with ambiguous messages with the result that many parents retreat physically from their children, thereby inclosing them in atrocious feelings of abandonment, loneliness and despair.

As a result of 1960s American pediatrics that advocated physical separation between parents and child which in the meantime is seen as a fundamental error, many of today's parents had a deprivatory childhood themselves and became dysfunctional parents of their own children. A long-term bestseller on the list of child-torture books from this generation of misguided pediatrics is the parenting manual by Dr.

THE DESTRUCTION OF THE NATURAL ORDER

Benjamin Spock that is still today a leading guide for many parents—to the detriment of their children.

In the contrary, it is through the absence of the father together with a shame-based identification in the mother-son relation that mother-son codependence is brought about. The reason for the more dramatic constellation in the mother-son relation has to do with the greater psychic fragility of the human male in general, and with the simple fact that it's the mother who is the matrix, not the father, in particular. If we want to add one more problem complex, which I did not research, it's the codependent mother-daughter relation.

By contrast, father-child care in our culture is seldom codependent simply because the father is most of the time absent. And this is, then, also one of the causative factors in mother-son codependence. But apart from this, there are singular cases of father-daughter codependence and they are marked by the fact that the father exceedingly overprotects the girl-child to an extent to virtually keep her 'away from life'.

As I have seen it in some families, this can bring about absurd constellations and relationships that symbolically express that the child is no more allowed to walk on their own feet, but on the feet of the father, so as to be 'protected of the harshness of life.'

The problem is much more manifest in white Western culture than in any of non-Western and tribal (native) cultures. My research has shown that virtually the only cultures *that do not have the problem* are tribal cultures, that is most native populations around the world. One important element in this etiology that has hardly been elucidated by research is that these children experience terrible loneliness during their childhood and youth.

The Pitfall of Emotional Entanglement

Another important insight about mother-child co-dependence is that it deprives the child, typically the boy, of the time and care needed for developing his true and individual intelligence, his own intrinsic gifts and talents.

Men who grow up entangled with their mothers are caught in a net of stiffening responsibilities, or obligations, or what is felt as such, which impedes them from really thinking of *themselves*, and minding their own business. The result is that they hardly think

THE DESTRUCTION OF THE NATURAL ORDER

their projects through to the end, having no time and rest for vision-building, constantly harassed by their demanding mothers, threatened as they are with love denial or even financial starving in case they disobey and begin to live their own lives.

In this sense, it can be said that the *son bears the cross for the sins committed by his mother*, and it's really a capital sin to suffocate a young man's energies and intelligence by throwing one's weight around as a mother and disregarding his fragility as a man.

In this sense, many women in our society need to be educated what right motherhood is about, and even more so, what *wrong* motherhood is. Not only is the *Oedipussi*—title of a movie by the German humorist, author and filmmaker *Loriot*—subject to ridicule and humiliation, he is also one of the major actors on the stage of child-focused sexual crime. Our mass media depict the truth in a distorted manner, suggesting with their politically correct rhetoric a boy had to care for his mom eternally, if he's a 'good boy.' These views have to be judged perverse, as they are really putting nature upside down. Childhood is transitory. Period.

NATURAL ORDER

The French child psychoanalyst and therapist Françoise Dolto has analyzed this problem in the mother-son relation, in her book *Psychoanalysis and Pediatrics (1971)*, and she writes:

> There are boys who stay lovingly fixated upon their mothers; their behavior is characterized by the fact that they do not attempt to 'seduce' any other woman. If the father is alive, the two men are constantly disputing, for the fact that the boy does not detach himself from his mother and searches out other love and sex objects proves that the boy has not liquidated - in a friendship of equality with his father - his pre-oedipal homosexuality. He will therefore prepare for getting 'in trouble' with his father through his difficult and provocative behavior.
>
> —Françoise Dolto, Psychanalyse et Pédiatrie (1971), p. 88. (Translation mine).

When the father has left and the young boy 'dedicates himself' to his mother, this behavior can be accompanied by real social sublimations, which are associated with the activities derived from the repression of genital and procreative sexuality, but this boy cannot behave sexually and affectively like an adult. He suffers from inferiority feelings toward men that he unconsciously identifies with his father; he can also be a hyper-genital who is always avid to get new sex partners toward whom he will never build real attachment, but he will show impotent in relations with any woman he really loves, because this is associated in his unconscious with the tabooed incestuous object. The messages those boys and young men are typically bombarded with by their mothers are, for example:

- You are egoistic

- You are like your father …

- Think a little of your mother …

- I'm always sitting at home, can't you make time and show me around a little?

- You should have gratitude for your mother …

- etc.

And when the boy is on the right track and really develops a unique genuine interest, mother will have enough reasons to tell him that he's inadequate for it:

- Why do you spend so much time for this, it leads nowhere?

- Stay with your feet on the ground, you have grandiose ideas …

- Like your father, big mouth and little essence …

- Others have done that before you, so where's the sense of it?

- You better spend your time taking care of your old mother!

- Why don't you follow my advice, you are just stubborn!?

- I always told you, but you know better …

Much evil in the world done by men has its roots here, in a stiffening mother-son relation that deprived the boy for years of his vital energies, blocking his *emotional flow* to a point of self-forgetfulness. This is, then, the reason why these men one day explode, so to speak, for thinking of themselves *for one time*, and do something horrible to a woman, a little girl, or an elder. And who goes to jail is always the boy, then a man, and not his mother. And that, in my humble opinion, should be changed. Women are to be made responsible for being abusive as mothers, not only men, as fathers!

Women always claim to not being given enough responsibility under patriarchy, but most women

THE DESTRUCTION OF THE NATURAL ORDER

bluntly deny their abusive attitudes toward their sons in our society, which is an abuse of responsibility, an abuse of power. However, this abuse is hidden for the most part, and often veiled behind feminist activism, a career or what I came to call a *victim attitude*. Women always cry for abuse when it's about *them*, never when it's about the sons they drive into madness, suicide, child rape or even murder. And here our laws have to change, definitely!

Of course, in clinical and psychotherapeutic practice, codependence does not in the first place manifest as a parent-child problem, but as a husband-spouse problem, and that is why it comes up in marriage counseling and family therapy. And that is exactly what makes it so intricate and difficult to heal it in the therapeutic setting.

What many practitioners overlook is that the problem *does not originate in the partner relation* but in the earlier parent-child relations that both partners experienced and that they project, as a matter of unconscious automatisms, upon their partner. We all project our parent of the opposite sex upon our spouse or husband, only that there are two essentially different ways of doing that, a conscious way based on the letting-go of the parent (mourning), or an unconscious way based on entanglement, confusion and hate-love.

In the Freudian terminology of the *Oedipus Complex*, the first alternative corresponds to what Freud called a liquidated Oedipus and the second corresponds to what Freud called an unresolved Oedipus.

Emotional Abuse

Introduction

Christopher Bagley writes in his book *Child Abusers: Research and Treatment (2003)*:

> Emotional abuse causes the most long-term harm to children, although combinations of emotional with physical and/or sexual abuse cause the most harm to long-term mental health.

What is *emotional abuse, emotional incest* or *covert incest*? I think that today many men have a quite sadistic relationship with women, which is something like a revenge reaction or compensation for the co-dependence they went through with their

THE DESTRUCTION OF THE NATURAL ORDER

emotionally abusive mothers. Unconsciously, they want to punish their mothers for the constant humiliations, the constant withdrawal of affection, the conditioned love they received and the painful lack of autonomy that is the sad reality in this kind of exclusive relationships.

The main problem in our culture is the mother-son relation and as good as all our social and relational problems flow out from this major distortion. Many men project their early ambiguous feelings toward their mothers later on their spouses, girlfriends, and even little girls they encounter, with the result that the ambivalent and hardly conscious aggression they foster toward their mothers is projected outward in society, and creates havoc in man-woman and man-girl relationships. This aggression in men comes about through the combination of lacking autonomy in their boyhood, absence of the father, demanding attitude of the mother for the son to stay at home, strict education with frequent humiliating punishment, isolation from peers through motherly overprotection, attitude to enclose the boy in an exclusive, intimate and emotionally abusive relation, victim attitude of the mother, and the explicit or hidden demonization of the boy's peer relations, friendships and social life.

A way out could be a certain persistence of the boy in the face of such a situation, and a firmness to be developed on his part that insists on his right to maintain relationships with peers, teens and adults other than tutelary figures and family, and that he asks for a certain amount of free time, every weekend, for going out alone, and unmonitored.

This could give the young male the opportunity to speak about his emotional pressures, about the humiliation he suffers and his confused feelings, especially when the boy turns into adolescence and these feelings of aggression start to get sexualized and become more or less violent sexual urges. While generally, with overprotected youngsters, a problem of acceptance will occur at the beginning in any group relation and a certain hostility may be experienced at the start, it can only be beneficial for young people to leave their nest from time to time to search out peer company and also adult males and females, who may be in state to support the young boy in his rightful quest for autonomy and respect.

THE DESTRUCTION OF THE NATURAL ORDER

The advice that I give for such cases is to strengthen personal autonomy, and to get into an inner dialogue with the shadow, and the inner child, in order to unveil the hidden distortions in the mother-son relation that has been internalized and that can be gradually rendered conscious through this kind of work.

The result of my more than thirty years of research on abuse and sexual paraphilias is that these sexual distortions result from mother-son codependence *that has reached a level of gravity to be qualified as emotional abuse*, and which is to be seen as one of the biggest relational problems of our times.

The Primary Abuse Etiology

Unfortunately, Western psychiatry only very recently began to get a hint of this, and when I started my research, back in 1985, there was not yet any book published on emotional abuse. Emotional abuse is now considered as the worst and most long-term form of abuse, as it's of all abuse etiologies the *primary etiology*. Sexual abuse is only one of several consequences of emotional abuse. Emotional abuse has become something like my research specialty and even now, I discovered, only very few books are published about it, while whole libraries have been written about sexual abuse. Contrary to most psychiatrists, I came to believe that the long-term psychic strain and fixations sexual abuse causes is not typically related to the sexual experience, if there was any, but to the following factors that are, or are not, present in such cases:

- Suddenness of the experience or child was trapped;

- Behavior was conflicting with the social code or family attitude;

- Entrapment effect that led to immediate anxiety;

- Debasing attitude of the type 'I can have all females I want';

- Impossibility after the experience to talk to anybody about it.

Much could be changed socially if anti-abuse social work could be based on these research insights instead of going on with tearing in the dirt pedoerotic sexuality, as this is the common public rhetoric in today's postmodern international

THE DESTRUCTION OF THE NATURAL ORDER

consumer culture. Here the focus is obviously wrong, as authors such as Stevi Jackson, a feminist activist, and Alayne Yates, an American child psychologist, have shown in their books.

—Stevi Jackson, Childhood and Sexuality (1982) and Alayne Yates, Sex Without Shame (1978).

The focus must be on fighting coercion, violence, and entrapment, not sexuality, and Western society should eventually learn to accept all forms of sexual behavior as a non-vulgar, non-harmful, non-debasing and creative human activity. Sexuality, after all, is a form of communication, and it's a social, not an asocial activity.

What Western culture does is to distort and pervert children's emotional life virtually from the cradle, and the Freudian myth of the *Oedipus Complex* has pretty much contributed to this distortion of the natural psychosexual growth of the child.

Children do not grow through being *codependent ersatz partners* of their parents, and yet this is exactly what the present culture is doing with them, imprisoning them in the nuclear family and depriving them of the whole bunch of hairy folk they were hitherto exposed to, when still living in the extended family and also a good part of the day in the street, without being constantly monitored and followed up.

The present structure virtually breeds violence, and this on a worldwide scale because the Western educational paradigm is exported all over the world within global consumer culture.

The Oedipal Mold

What means Oedipus Complex?

Sigmund Freud (1856–1939), an Austrian neurologist and co-founder of the psychoanalytic school of psychology believed that psychosexual growth comes about through three stages, the so-called oral phase (0-2 years), anal phase (2-4 years) and genital phase (4 to 7 years, followed by the latency period (7-11 years) and adolescence (11-16 years) and that the child invariably passes through these stages.

> —I have oversimplified the age groups here for easier understanding; in reality the border line between the phases is not that clear-cut. Most psychoanalysts, for example, let

THE DESTRUCTION OF THE NATURAL ORDER

the oral phase end with the age of 18 months of the infant, and put this also as the ideal end of the primary symbiosis with the mother. In addition, it has to be seen that children are not automatons and do not follow those schemes on the letter, which is especially true for highly gifted children. It is known, to give an example out of context, that the late pianist Arthur Rubinstein did not speak before the age of three, but he spoke at once several languages.

In addition, Freud argued that the intrinsic setup of the sexual drive structure was taking place through identifications, especially the identification, during the anal phase, with the parent of the same sex, that Freud called *homosexual identification* and the following *heterosexual identification* with the parent of the opposite sex, during the genital phase.

This latter sprocket in the psychosexual machine of sexual growth was called *Oedipus Complex* by Freud. More specifically Freud and later psychoanalysis require the child to successfully liquidate each phase or fixation, and conclude that if a child was not able to do such liquidation, the sexual energy would become stuck in the particular phase where the development was arrested, with poignant consequences on sexual habits.

For example it is argued that when a child does not successfully liquidate the *Oedipus Complex* by developing a strong heterosexual relationship with the parent of the opposite sex (without however acting this attraction out), the child would probably become homosexual later on. Freud has found this first for boys with regard to their mother, and later added it on for the girl-father relationship, which he called *Electra Complex*.

Is the Oedipus Complex Universal?

I think a number of intelligent and child-loving people find it makes sense when Freud affirmed the basic sexual nature of the child and infantile sexuality. But my question is if this understanding really implies that they see and acknowledge Western culture's fundamental denial of the child's affective, emotional and sexual complexity? As a parent, to allow one's child to be sexual in a culture that is outspokenly against that kind of freedom really is a challenge; that is why only when parents get the whole picture, they can do what needs to be done. If parents are wishy-washy on this question, and half-hearted, it makes it probably only worse.

When I was starting my research, I honestly had no idea that children could have an authentic sexual life, I mean in the sense of copulating with each other, and not just in the sense of being autoerotic through masturbation or mutual masturbation with a friend.

I learnt these facts through anthropological field work, through ethnological reports published by Bronislaw Malinowski, Margaret Mead, and others, and through literature on alternative childhood, and children in the counter-culture.

> See, for example, Bronislaw Malinowski, The Sexual Life of Savages in North West Melanesia (1929) and Sex and Repression in Savage Society (1927), Margaret Mead, Sex and Temperament in Three Primitive Societies (1935), Susanne Cho, Kindheit und Sexualität im Wandel der Kulturgeschichte (1983); Larry L. & Joan M. Constantine, Treasures of the Island (1976) and Where are the Kids? (1977); V. Elwin, The Muria and their Ghotul (1947); Richard L. Currier, Juvenile Sexuality in Global Perspective (1981), 9 ff.

In the absence of this knowledge, Freud's theory that children's psychosexual development was a process of libidinal identifications was for me an *attractive surrogate for the real knowledge!* And it is an attractive lie, for it *justifies the existence of the holy consumer family* with a child as the main stage clown who is used and abused under the pretext of his or her *needs* —while the reality is that this psychological construct rather serves the parents' needs for emotional security and the socially sanctified and legally imposed avoidance of children's real autonomy through real erotic experience with people outside of the nuclear family.

This reductionism is the pseudo-scientific cover-up of today's mainstream child psychology; it could appropriately be called child sex mythology! Freud was the avatar for what later became, and today still is, the mainstream paradigm in child psychology and education.

My research on codependence brought me to retracing the functional link between identity and autonomy, pointing straight away to the pitfalls in the Western educational system. One of these pitfalls is the exclusion of parameters that serve to build identity through self-knowledge, intuitive or inner knowledge, psychic knowledge, pre-life knowledge and relational experience.

The identity that is said to be the only possible mold according to Western mainstream psychiatry is a derived, not a genuine, identity. It is derived from the parents' identities. For a boy, for example, the process will be identification with the father, as a primary homosexual identification, during the anal phase and identification with the mother, as a secondary heterosexual identification during the genital phase.

According to Freud, the *Oedipal Complex* comes in at that moment in the child's psychosexual development. True identity is built, according to this theory, when the boy has successfully liquidated the *Oedipus Complex* by having developed *enough aggressiveness toward the father and enough castration of his incestuous desire toward the mother at the same time.*

That this system is built upon the grave of child sexuality, in the sense of child-child sexual activity, is clear from the start. It was clear to Freud but he thought that a deeper yielding to the core of nature's laws would catapult Western bourgeoisie into chaos.

I have critically reviewed Freud's theory of *infantile sexuality* and came to the conclusion that Freud's scheme is clearly detrimental to the child's building autonomy, by keeping the Western consumer child in pseudo-fusional dependence on their parents, thus creating codependence and perversion, and a fake heterosexuality that covers up all the undealt-with secondary drives that are produced by forcefully impeding the child from living out their natural emosexual attraction toward peers.

My wake up call finally came not from psychology, but from the side of anthropology and the insights I got through my studies of the human field, the energetic functionality of the organism and the nature of the bioenergy. It was first through the anthropological findings of Bronislaw Malinowski and Margaret Mead and their observations of biologically healthy child-child sexuality with the Melanesian Trobriand culture and other tribal cultures, that triggered a change in my regard on child sexuality.

We have two ways to create a new reality, in which society, recognizing the child's affective, emotional and sexual complexity and high bioenergetic charge, sets up new and comprehensive forms of child-rearing:

- by confining the child in an oedipal triangle within the nuclear family, depriving them of all and any non-incestuous erotic relations, and artificially raising their gerontophilic eroticism, while projecting this eroticism exclusively upon the parents, thereby creating a striking conflict within the child's psychosomatic setup;

or

THE DESTRUCTION OF THE NATURAL ORDER

> ▸ by transforming mainstream culture and granting children their own domain of intimacy, outside of the parent's embrace and control, allowing the child to live their affective, emotional and sexual complexity in freedom, thus helping the child to build true autonomy and self-reliance.

The first alternative leads to the consumer child. The second alternative leads to a complete human.

To summarize, Sigmund Freud has significantly contributed to consolidating what I call *Oedipal Culture*, to a point to have prepared the subtle ideological soil for the most sordid fascist ideology of humanity, postmodern international consumer culture.

Freud has less significantly contributed to helping the modern child with their *natural quest for autonomy and self-reliance*, and their birthright for an unobserved realm of intimacy, outside of the jovially persecutory parental and educational embrace, if not to be kept save from the Kindergarten regime of slave-puppets to their culturally perverted and schizoid parents and educators.

Criticism of the Theory

1/8

The Freudian scheme is only if ever valid for cultures where child-child erotic relations are forbidden and structurally impaired, that is, for patriarchal culture and postmodern international consumer culture as the successor, in a new garment, of the patriarchal rut;

2/8

The Freudian scheme represents systematic perversion of the child and implies the cultural conditioning into homosexuality because identification is not the natural way for the child to build their love map, and to individuate, but a *culturally conditioned one*, which is why I call this kind of culture also *Hero Culture,* implying the child is molded after their parents taken as cultural standard models, and not in relation to their own specific soul structure, content and emotional setup.

3/8

Building homosexual attraction before building heterosexual attraction is not the way nature builds

THE DESTRUCTION OF THE NATURAL ORDER

our psychosexual structure, but is a pure projection upon nature. Small boys are erotically attracted to their mothers and girls to their fathers, and not homosexually toward their same-sex parents. This is so from birth, not just from age four or five, as the Freudian myth assumes. When, as this is admittedly often the case within patriarchal cultures, children are homosexually fixated upon their same-sex parent, and refuse to open up for embracing their parent of the opposite sex, exhibiting anxiety in front of anything erotic, this is so because the child is *narcissistic and neurotic*. Needless to add that the neurotic child is of course not the natural child; when this happens, it has a reason, as it does not happen with children who are educated with love. I have personally seen it over and over with children whose parent of the same sex gives only conditioned love, and where children lack emotional constancy and security with their parents, or even grow in disruptive and dysfunctional families.

4/8

Freud's professional and private life philosophy was patriarchal and at the same time materialistic, and mechanistic. He had discarded out of his life any spirituality as well as his wistful Jewish tradition, and most of his theories defy truly spiritual insights and truths. The very essence of a *holistic worldview* that sees the hidden connections, was alien to Freud. This was one of the reasons that his relational life was full of strife and disruption, and ended in many painful separations and personal conflicts. It can be said, *cum grano salis*, that Freud was abandoned later in his life by all his real friends, which was *one element* in the etiology of his atrociously painful death of jaw cancer. When we consider that we are talking about love and erotic attraction, when we talk about the psychosexual growth of the child, it denotes confusion to choose Freud as the wistful authority on the matter. He surely was not. That Freud's theories are slavishly followed till today has *political reasons*, and is in no way to attribute to any real insights he had. In fact, Freud's psychosexual theories are the *ideological base justification for the enslavement of the consumer child*, with all that results from this cultural perversion.

5/8

    Freud overlooked not only female sexuality, as the feminist movement alleges, but he also overlooked that the small child is not an autoerotic freak and serial masturbator when being allowed to have full relations with other children. Freud ignored the real natural emotional and sexual growth processes in children, as they are amply demonstrated by non-patriarchal cultures where children enjoy full sexual freedom from early childhood. In these cultures, children engage in sexual peer relations that are tolerated and encouraged, by not interfered with by tutelary adults, as shown by the ample research of Bronislaw Malinowski, Margaret Mead and Wilhelm Reich, and many others.

6/8

Freud's theory of *infantile sexuality* reflects the power structures of patriarchal society; he just put names on things that were already there. In fact, today's global consumer society is unthinkable without the dogma of the *Oedipus Complex*, the resulting parent-child *codependence* and the confusion it brings about in the mind of the child. When natural peer relations are forbidden to the child because autonomy and self-thinking abilities of the child are replaced to a large extent by *system-conform consumer conditioning*, the way is open for total addiction in form of non-ending consumption. The result is the perverse consumer child, and the so-called citizen, that are both based on the massacre of the original primal child that was naturally heterosexual—and more generally so, *sexual* in the first place.

7/8

Freud's theory of the *polymorphously perverse infant* is a direct result of the mechanistic science tradition along the lines of Jean-Jacques Rousseau, Isaac Newton, La Mettrie, Baron d'Holbach, René Descartes and others, which considered man being a machine and infants to be born as a *tabula rasa*. While this view today is scientifically outdated and while we know that infants are born with a full heritage of former incarnations and resulting imprints in the soul, positivistic child psychology has to this day not done the necessary shift from a blind mechanistic and highly doctrinaire pseudo-science into a real holistic science of the bioenergy. It assumes that our emotional identity is a soul imprint, which is the blueprint of our later individuality. It also assumes that all in life is a function of the *human energy field* or *quantum vacuum*. Hence, sexuality is but flowing vital energy streams and has very little to do with the mechanistic assumptions an ignorant sexology and a myth-ridden psychology projected upon it.

8/8

Responsible parents raise their children in total opposition to Freud and the cultural slavery that his theories and the power structures of patriarchal society require, and give their children ample opportunity for peer-peer, and peer-adult, emotional and sexual relations, by interfering as little as possible in their children's love lives, which includes avoiding both emotional and sexual incest, while at the same time encouraging the child to project their libido on figures outside of the family framework.

OEDIPAL CULTURE

My critique of *Oedipal Culture* is inextricably woven with my criticism of Sigmund Freud's *cultural concept* of psychoanalysis, and here especially my critique of his theory of the *Oedipus Complex*.

Many young parents believe that psychoanalysis had contributed to the liberation of the child; they tend to think it was a professional vintage of *permissiveness*, or a variant of permissive education. Nothing could be farther from the truth. Freudian psychoanalysis, applied to children is not permissive, it is *normative*; it is a tool for forging the ideal consumer child within consumer culture that is based on the economic paradigm of total consumption. As such, it is an ideological pillar for the functioning of a society that needs to repress natural pleasure because it replaces it by *consumer pleasure*.

Psychoanalysis is not permissive at all. It can be proven statistically that the word most used in psychoanalytic publications is the word *castration*. Castration is a highly violent term that suggests the cutting off of the male sexual organ or the infibulation of the female sexual organ, the latter often also being called *clitoridectomy*. While psychoanalysis claims to use a *mythical or metaphorical vocabulary*, this vocabulary becomes strangely real when it goes to take a measure that will affect the long-term destiny of a child or a family.

In discarding out children who are judged as *sex offenders* or social delinquents, psychoanalysis exerts its full social power in that it can put people, not only adults, but also children, in jail. The children's jails are cutely called educational rehabilitation centers, but their regulating principles are the same as those of jails for adults, however with the difference that in child jails to this very day constitutional guarantees are absent, while those guarantees are well in place for adult prisons. This shows, more than anything else, the true attitude of *Oedipal Culture* toward children, as it shows the devil's face of this matter called *child protection*.

Are Masturbating Children Better Citizens?

Françoise Dolto, the late French child therapist and psychoanalyst is very outspoken about the benefits of masturbation but we are not set in the world to masturbate, but to copulate and lovingly embrace others. We are not set in the world to engage in endless autoerotic self-satisfaction, but to use our natural erotic desire for building *relationships*. In this sense, sexuality is social, a social factor, and social behavior. Hence, people who are sexual are more social than those who repress their sexual wishes.

Child development, as a whole, cunningly cheats about this fact and relegates the child to eternal masturbation in the name of their own best. Children are encouraged to develop the habit of masturbation, instead of learning to make love with another human, which is the real, and natural, form of loving sexual embrace.

What a split paradigm this is! The child is encouraged to be autoerotic and to develop erotic fixations upon their parents, but violently, with all the police power in modern society, withheld from engaging in what is most natural: to embrace others lovingly, others who are not incestuous objects, and thus peer children and adults other than their parents.

Western culture's child-rearing paradigm, whatever Dolto and others had and have to say about it, is perverse in my view, as it really puts life upside down in the name of culture, morality or whatever other fake arguments.

Dolto encourages professionals to take note of the child's sexuality to better serve the child, but what is this service about down the road? To transform loving children into egoistic masturbators and incestuously fixated psychopaths? The functional organic troubles she mentions in her books are often the result of *love prohibitions*, not prohibitions to masturbate, but prohibitions to have real love relations outside of the family, and to have the basic freedom to build such love relations in the first place. See the following quote:

> FRANÇOISE DOLTO
>
> All those who study behavior problems, functional organic troubles, the educators, the doctors in the true sense of the term, must have notions about the role of libidinal life and know that sexual education is the grain for the social adaptation of the individual.
>
> —Françoise Dolto, Psychanalyse et Pédiatrie (1971), p. 63 (Translation mine).

It is of course true what Dolto says about the negative effects of prohibiting masturbation. But the trick is that the reverse argumentation is not per se correct. To allow masturbation does not mean to give the child *real freedom* for love. This is the logic error here, and here is where society cheats the child and argues from an irrational and mystical position that is not factually verifiable. The prototype example for this mysticism is where society or psychoanalysis—and here they lovingly coincide in their spanking the consumer child—speak about *pedophilia* when the question is not giving pedophiles their right, but giving children their right to love adults. These are two different matters, do what you will, but they are thrown in one pot and judged as one and the same thing. Here is exactly where the trail of lies begins.

> Françoise Dolto
>
> To prohibit the child to masturbate and sexual curiosity means to force the child to pay unnecessary attention to activities and which normally, before puberty, are unconscious or preconscious. (…) Developing consciousness prematurely in an atmosphere of guilt does great harm to the development of the child because it deprives the child of ways to use their vital energies (libido) that is inherent in those spontaneous activities. Psychically healthy children who have mastered the genital stage are toilet-trained, graceful in their body and dexterous with their hands, they talk well, listen and observe a lot, like to imitate what they see others doing, ask questions and expect truthful answers, and when they don't receive them, begin to make up magical explanations.
>
> The truth is that normal masturbation does not at all fatigue the child, but appeases the phallic vital tension of which give his erections ample evidence. Masturbation provides the child with physiological and affective relaxation which does not equal in intensity the orgasm of an adult as there is no ejaculation (…). (Id., p. 66, Translation mine)

THE DESTRUCTION OF THE NATURAL ORDER

The Dogma of the Autoerotic Consumer Child

It goes without saying that for those who are against all expressions of children's eroticism, Dolto's ideas about *child masturbation* will probably sound somewhat progressive or permissive. But from the background of the larger picture that I am trying to paint here, masturbation, while it's good of course and while many people, big or small, need it just for getting rid of their surplus bioenergetic charge, is not the real thing what the child needs and asks for.

To repeat it, we are born to learn copulating, not masturbating, and what children should learn instead of becoming proud masturbators is to become *humble partners in a real sexual embrace* where set and setting are correct, and where there is mutual respect, dignity, love and acceptance. To say this, excuse me, is not an apology for *pedophilia*, as such a social policy, once enacted, would naturally lead, just as in most native cultures, to sexual relations among children.

If a random number of children choose adult mates, this then has to be respected, for there can only be *one* result when we give the child the right for free choice relations. If children are free to choose their mates, they must be allowed to engage with adult partners as well. To do so does not imply a legal implementation of pedophilia as a new social and legal paradigm, and I am very explicit about this! However, it well implies that there is no criminal punishment for adults who engage in sexual relations with consenting children.

But as matters are in our culture, the basic resistance against children as erotic beings is primarily child-child sexual interaction. According to Freud's cultural preservation theory, to admit and endorse child-child sexual relations is against the setup of our culture. This dogmatic position of Freud is documented and led to a number of conflicts with his students. It was the main reason for Wilhelm Reich taking a distance to Freud, after the latter said regarding Reich's activism for the sexual liberation of children 'Culture must prevail!'

Françoise Dolto, when I interviewed her in 1986 in Paris, put it in the following terms:

> It is true that Freud was normative in this matter. But why not? The task of psychoanalysis is not to trigger a social revolution or changing the cultural paradigm. We are

THE DESTRUCTION OF THE NATURAL ORDER

here as psychoanalysts to heal the neurosis, in the individual case, that comes from the cultural repression of the child's sexuality. This is our task, not more and not less. Freud has seen it in the same way. (Quoted from memory, Translation mine)

Hordes of psychoanalysts followed their master guru in this greatest myth of all myths that Freud created with the whole of his doctrine of the *Oedipus Complex*. It may be against our tradition to eventually accept the child's full sexual freedom, but *every culture can change,* and only when it's in constant change, it's alive. A culture that never changes is a dead culture, and a dead culture is a no-culture.

In truth, what Freud ordained here as some kind of cultural imperative was a command to uphold patriarchy; he cannot be taken as the progressive child-loving psychoanalyst that history has made out of him, but a reactionary! His 'Oedipal' doctrine is a recipe for cultural neurosis and stagnation, not for cultural progress. The advice Françoise Dolto gives to parents for the child who is found to masturbate often is equally ambiguous, and suspiciously on the line of Freud's cultural reasoning.

She argues such a child would have to be *initiated*. Until here I agree. But she continues that such a child has to be initiated into superior activities, which require a *higher mental level* than those usually reserved for children of that age.

> FRANÇOISE DOLTO
>
> [W]hen you see a child masturbating often, a child who is normal, you can be certain it's a gifted child that should be initiated into superior activities, which require a higher mental level than those usually reserved for children of that age. But even more often, it's a neurotic child for whom masturbation has become an obsessional habit. Such a child must be given treatment, not punishment. To intimidate the child, or even prohibit masturbation will impair the development of the child; in case the child obeys the prohibition he will become dull and insensitive, and if he does not obey he will become unstable, angry, undisciplined and revolted. Neither of this is intended to be brought about by the adults who react in those ways; but this is what adults are doing to children, without knowing what they are doing. (Id., p. 74)

INTELLECT BOOSTING FOR SEXUALLY DEMANDING CHILDREN

That means a child who is longing for stronger sexual fulfillment than that of masturbation has to receive an intellect boost. That is really giving a child a pear who asks for an apple. What a child naturally wants is to be initiated into loving copulation,

THE DESTRUCTION OF THE NATURAL ORDER

because in masturbation, as all my research on the human energy field clearly shows, the vital energy level is well brought to a new balance through orgasm, but that is not all there is in sexual love. What is perhaps even more essential than the sexual abreaction is the *tactile experience* of two nude bodies being close in excitation for a while, which namely results in a high-level exchange of bioelectricity and *emotional flow* which is like feeding our internal batteries, strengthening our immune system and working counter to the aging process.

From this larger picture that I tried to paint here, the pretended revolution of so-called *infantile sexuality* sounds like a bad joke, if it was not a bad trick, and actually a big lie and a real enslavement of the child in the name of a life-denying dead culture that knows only to consume and to possess, and as a result, to conquer and to rape, but not to live and to love and respectfully embrace.

Of course, what Dolto reasons here on the development of the rational mind is all true; it's genitality that brings about the objective mind. But our society is not a group of genitally developed individuals, which is why it is so deeply irrational and mystical, and so little responsible. Our society is one of anally fixated fabulators who are caught in the trap of mysticism that they call, in their madness, *psychoanalysis*. To take an ideological crap science and culture-protection system such as psychoanalysis for the ultimate truth about life or childhood is about the greatest madness I have ever heard of in my life.

What Dolto says in the following quotes is valid even more for real genital cultures such as the Trobriand islands where children learn to copulate from early age, and not, as in our culture, to become virtuous masturbators and pleasing night cushions for their emotionally frigid parents. But the difference is that they do not need the whole of the Oedipal construct, with its detour to arrive at genitality and heterosexuality via homosexuality, simply because they give *real freedom* to their children, and real sexuality, not a perverted form of it. And that is why the outcome is *real heterosexuality*, and not, as in our culture, fake heterosexuality.

> FRANÇOISE DOLTO
>
> It is only after the liquidation of the Oedipus that thought can be put at the service of so-called altruistic sexuality, which means that seeking narcissistic satisfactions must have been overcome, without

however invalidating those satisfactions. In the genital state, thought is characterized by common sense, prudence, and objective observation. It's what we call rational thought. (Id., p. 54, Translation mine)

Qualifying Oedipal Castration as Child Abuse?

My criticism of Dolto's approach to child sexuality, as I was on good terms with her and exchanged with her for a while, may sound strange and exaggerated, but it is not in any way directed against her personally. I am speaking here about the *perversity of the whole of psychoanalysis*, the whole theater and comedy it represents, the grotesque family scenarios it plans and puts on stage, and the whole abstruse worldview it embodies.

What Dolto explains in the following quotes is certainly true, sadly true, as it exactly shows the shadow side of the whole of the Oedipal construct, and what it results in when the boy does not make it to *liquidate his Oedipus*, as psychoanalysts express it. And yes, the problem is more stringent with boys than with girls, for reasons we do not yet fully understand, but it has been argued by many psychologists that

NATURAL ORDER

men generally are psychically more fragile than women.

FRANÇOISE DOLTO

There are boys who stay lovingly fixated upon their mothers; their behavior is characterized by the fact that they do not attempt to 'seduce' any other woman. If the father is alive, the two men are constantly disputing, for the fact that the boy does not detach himself from his mother and searches out other love and sex objects proves that the boy has not liquidated—in a friendship of equality with his father—his pre-oedipal homosexuality. He will therefore prepare for getting 'in trouble' with his father through his difficult and provocative behavior. (Id., p. 88, Translation mine)

When the father has left and the boy 'dedicates himself' to his mother, this behavior can be accompanied by real social sublimations, which are associated with the activities derived from the repression of genital and procreative sexuality, but this boy cannot behave sexually and affectively like an adult. He suffers from inferiority feelings toward men that he unconsciously identifies with his father; he can also be a hyper-genital who is always avid to get new sex partners toward whom he will never build real attachment, but he will show impotent in relations with any woman he really loves, because this is associated in his unconscious with the tabooed incestuous object. (Id., pp. 88-89, Translation mine)

This is how the superego of the boy becomes very early rigid (...); the reason for this is the necessity to repress

THE DESTRUCTION OF THE NATURAL ORDER

the heterosexual desire in the 'maternal sphere.' (Id., p. 89, Translation mine)

The symbiotic fixation on a parent, especially the mother, beyond the natural mother-infant symbiosis, and thus after the age of 18 months of the infant, is pathological and it brings about a clear reduction of intelligence because of the entanglement of the vital energies of parent and child. This is particularly true, as Dolto points it out, in the mother-son relation, and less in the father-daughter relation because the mother-matrix has naturally a greater attraction power for the child than the father-spermgiver.

When mothers do not encourage their children to develop autonomy, they are on the best way to entangle their children in a codependence where the parent is the winner and the child the loser, and where the child, in most cases without parent and child being really conscious of it, becomes the *ersatz-mate for the parent*. While this mating is in most cases not sexual, the consequences of mother-son codependence are devastating.

I talk about *emotional abuse* in cases where the parent has received clear signals from the child for being granted more freedom and autonomy, but does repeatedly not comply with this request, or even actively cuts down or prohibits love and erotic relations of the child with persons outside of the family, whatever their age. Last not least, it doesn't come as a surprise when Dolto categorically judges perverse behavior and social delinquency as the result of a non-liquidated Oedipus, or one that is not yet liquidated.

> Françoise Dolto
>
> [P]erverse behavior or social delinquents, both are the result of a non-liquidated Oedipus, or a not yet liquidated one. (Id., p. 130, Translation mine)

Rationality vs. Oedipal Mysticism

The judgmental attitude of psychoanalysis is not surprising; it rather shows how devastating the Oedipal construct is at the end of the day, together with all the cultural weed that has grown around it. This insight, that is shared by most psychoanalysts and psychiatrists is not the real bomb; the real bomb is the fact that our society tolerates psychiatric nonsense that perverts our children into potential violent perpetrators, using a construct for the psychosexual growth of our children that is anti-life, dysfunctional, dangerous and unnatural.

THE DESTRUCTION OF THE NATURAL ORDER

There must be an awakening one day, and perhaps a movement is to be created that is similar to *Antipsychiatry* in that it clearly unveils the social utilitarianism of Oedipal Culture's child development paradigm because what it creates is not psychic health and responsible citizens but *emotional and sexual cripples* and a horde of silent anarchists who, while paying lip service to order and morality, are in fact *barbarous uncivilized rapists* because they have never learnt to copulate and embrace another in love when they were young and still open for sexual learning.

Modern rape research has shown that rapists are highly sexually inexperienced individuals who foster in most cases a *repressive* and *moralistic* worldview. These people suffer not from too much but from *too little permissiveness* and a blown-up super ego, and they are usually endorsing educational violence. In addition it has been shown that they are hostile toward healthy and caring touch, and suffer from actual tactile deprivation.

It is for this reason correct when researchers on the roots of violence, such as Dr. James W. Prescott, suggest to treat sex offenders with sexual permissiveness, granting them relaxation, massage, psychotherapy and frequent loving sexual embrace.

> —James W. Prescott, Body Pleasure and the Origins of Violence (1975) and Deprivation of Physical Affection as a Primary Process in the Development of Physical Violence (1979), pp. 77, 78.

Seen from a social policy point of view, we must conclude that it's exactly this denial of *real child sexuality* in the form of an active involvement of children in love relations outside of the family that renders our culture so outright false, morally corrupt, violent and destructive. And what we get from the pulpit of psychoanalysis here is but reject and denial, a false, jovial and grinning pseudo permissiveness which is an outright betrayal of the child, together with cathedral lectures from a blown-up patriarchal superego incarnated in women like Dolto, who 'speak the rude truth in all ways', to paraphrase Emerson.

Only that contrary to Emerson's, this truth does not liberate, but enchains our children in still more co-dependence, still more emotional entanglement and abuse and still more murderous fascist ideologies to come from this soil of a deeply perverted psychosexual base structure, which is the rotten foundation of our culture.

THE DESTRUCTION OF THE NATURAL ORDER

OEDIPAL HERO

Oedipal Hero is a term I have forged for an individual, usually of male sex, who suffers from a specific pathology that comes from a combination of an unresolved *Oedipus Complex* and a narcissistic fixation. In my view, modern psychiatry has just begun to identify this problem, and my approach to scientifically and psychologically outline this pathology is therefore to be seen as a pioneering work.

I use the term *Oedipal Culture* or *Oedipal Consciousness* synonymously with a range of similar expressions so as to denote the complex process of *denial of truth* about the cyclic and pleasure-bound nature of life through the repression of the child's emonic vitality.

> WILHELM REICH
>
> The unarmored organism does not know an impulse to rape and murder little girls, or to get pleasure through violence. It is therefore indifferent toward all moral rules that try to repress such impulses. It cannot comprehend that one has intercourse with another only because there is an opportunity for it, for example being in one and the same room with a person of the other sex. The armored character, by contrast, cannot envision an orderly life without coercive laws against rape and lust murder.

NATURAL ORDER

—Wilhelm Reich, Ether, God and Devil (1972).

While the true reason for repressing the child's vitality is hardly ever discussed in international consumer culture, the lifting of the veil behind so-called morality used to be a strong domain of post-revolution French philosophy. Most people in modern consumer culture really believe the main reason for inhibiting the child's free sexuality had to do with morality or with a concern for protecting the child's natural vulnerability. This cultural and social naiveté strongly contrasts with other cultures' perspective, such as the French or Hispanic cultures, and it stringently contradicts the life and love philosophy of most tribal cultures.

French social historians such as Michel Foucault and social philosophers such as Gilles Deleuze or Felix Guattari have clearly demonstrated that the reasons for the child's emotional castration are to be found in the setup of Western consumer economy. It has *economic*, not moral reasons why the Western consumer child is relegated to *forced orality* and deprived of tactile stimulation.

Gilles Deleuze and Felix Guattari, in their philosophical exposé *Anti-Oedipus, Capitalism &*

THE DESTRUCTION OF THE NATURAL ORDER

Schizophrenia, set out to formulate a fine and detailed philosophical, logical and ethical critique of Freud's theory of the *Oedipus Complex*.

—Gilles Deleuze, Felix Guattari, L'Anti-Oedipe (1973).

To illustrate my own point of view, subject to several of my books, I will provide here some quotes of this major philosophical and psychoanalytic treatise. All the quotes are taken from my own translation of the French original.

> GILLES DELEUZE, FELIX GUATTARI
>
> People often believe that with Oedipus, it's easy, and you can take that for granted. But it is not so: Oedipus presupposes an extraordinary repression of desiring machines. And why, and for what reason? (Id., p. 8)
>
> GILLES DELEUZE, FELIX GUATTARI
>
> Does Oedipal imperialism only require to abandon biological realism? Or has something else, infinitely more powerful, been sacrificed to Oedipus? (Id., p. 63)
>
> GILLES DELEUZE, FELIX GUATTARI
>
> The un-Oedipal nature of desire production continues to exist, but is aligned with Oedipal coordinates that translate it in 'pre-Oedipal,' 'para-Oedipal' or 'quasi-Oedipal,' etc. (Id., p. 65)

NATURAL ORDER

Mysticism and Atheism

Scientific Mysticism

Most people think Western science was characterized by a definitely *rational* approach to reality. Contrary to the natural order, they consider modern politics as rational and ethical policy making as a rational approach to leading people, countries and international organizations.

Atheism can be a fruitful transition from religious indoctrination to true liberal spirituality, but when it's a lifetime business, it takes the form of knowledge denial, and then it's pure reductionism and ignorance in the form of 'I see my body is flesh and bones only and that it decays after death, so how can there be anything beyond the reality of our senses? If there was anything, I could feel and touch it.'

Atheism is an absolutism of the five senses, a reduction of the wholeness, vastness and completeness of life to a *residue concept* that is but a tiny splitter of reality. And as such, it's first of all a *reduction of perception*, and in most cases a categorical exclusion of its most original vintage: *direct perception*. The process of perception is confounded with information processing, whereas of course both processes, that the brain does simultaneously, are functionally different and serve different purposes.

To put it in simple terms, and to paraphrase Wilhelm Reich, it's the expression of the human *No* in the face of the greater life, which is basically transpersonal and which includes other, and perhaps parallel, realities (for which we do not have built-in sensory perception abilities), and for the perception of which we would need to build and train our extrasensory, multisensory or metasensory perception potential.

What I am saying is that from a position of denial and fragmented perception, and from a paradigm of reality reduction to the known, as Krishnamurti put it, and the limited realm of sensory experience, it's but a step to end up in *mysticism* as a cover concept for intellectual anarchy and the supremacy and sanctification of spiritual ignorance.

Mystical Thinking vs. Functional Thinking

As Wilhelm Reich put it, mysticism truly is part of mental insanity, and its confusion with spirituality is a rather typical result of mechanistic thinking. Thus, the mechanist and hyper-rational functionary is only a step away from becoming a fanatic mystic. Wilhelm Reich writes in his report of a schizophrenia case:

WILHELM REICH

The schizophrenic world in its purest form is a mixture of mysticism and emotional inferno, of penetrating though distorted vision, of God and devil, of perverse sex and murderous morals, of sanity to the highest degree of genius and insanity to its deepest depth, welded into a single horrible experience.

—Wilhelm Reich, The Schizophrenic Split (1972), p. 1.

Wilhelm Reich was one of the rare functional thinkers of his era and who by the mass mind was invariably qualified as a mystic. This is so until this day. Even many of the self-labeled orgonomists and especially the hagiographers of Reich put the late German doctor, psychiatrist and bioenergy researcher up as a mystic genius who was largely misunderstood because of the latitude, depth and complexity of his mind.

This is simply not true. Reich was in no way a mystic, but a clear, rational, functional and yes, broad-minded thinker, a scientific mind *par excellence*. And because of the fundamental, and

THE DESTRUCTION OF THE NATURAL ORDER

all-so-typical, misunderstanding of this unique and highly gifted scientist, Reich had to become explicit in explaining *what mysticism really is—and what it is not*.

That is why I am going to use a number of quotes from his writings here to explain what mysticism really is, and that it has nothing to do with true spirituality. Reich writes in *Ether, God and Devil*:

> WILHELM REICH
>
> Functional thinking as a research principle of energetic functionalism is best seen at its work in understanding the unity of soul and body, of emotion and arousal, of sensation and stimulus. This unity or identity as a base attitude in observing the living clearly excludes otherworldliness, and even the idea of the autonomy of soul. Emotion and sensation are bound to physiological and orgonotic arousal. This excludes any form of mysticism, for mysticism is characterized by an otherworldly autonomy of emotion and sensation. This is why any regard upon nature that affirms the autonomy of soul live, whatever it is labeled, is mysticism.
>
> — Wilhelm Reich, Äther, Gott und Teufel (1983), pp. 95-96. Translation mine. This book was originally published in its English version: Ether, God and Devil, ©1949, 1972 by Mary Boyd Higgins as Trustee of the Wilhelm Reich Infant Trust Fund and published by Farrar, Straus & Giroux, New York. The first edition of Ether, God and Devil was published in the English language as Volume 2 of The Annuals of the Orgone Institute in 1949.

Reich explained in this quote that spirituality, if it is true and not just mysticism, must be based not upon belief, but upon real *spiritual knowledge*. At the time of Reich, it is true, knowledge about what soul reality means was scarce and it was entangled with indoctrinating and irrational religious thinking and maneuvering.

What is soul? Has soul ever been scientifically defined? It has, in the meantime. Soul may be an awkward expression, but we have today clear accounts of, and scientific evidence for, the afterlife, of the realm of spirits and other disincarnate entities, of psychic powers, of precognition and remote viewing, of the independence of life from the physical shell, and of parallel universes. It was mainly *quantum physics* that through its paradoxes lifted the veil of what formerly was held for mysticism and integrated this missing knowledge into a broader-defined science that is called holistic or meta-rational.

It may not be obvious for many people today why Reich had to explain so extensively why he was a *functional thinker* and not a mystic and sex-obsessed freak. It has to be seen that orgonomy as a science and, at its basis, the existence of the human energy field simply was denied in the West, and is even today is denied in other than avant-garde science circles. Reich experienced frequent attacks from groups close to churches, and they argued Reich's research on orgasmic streaming was but 'scientific' pornography. The following two quotes may serve to clarify this issue:

> WILHELM REICH
>
> Research in the nature of sensation led to the theoretical, practical and experimental discovery of physical orgone energy that possesses specific biological functions. This discovery could never have been made by a mystical worldview that affirms soul motion. This is a matter of principle, because the mystic does not see the connectedness of soul and body, and this is also a practical matter because mystical thinkers do not acknowledge their organ sensations, or they experience these sensations in a distorted way, and not in an immediate connectedness like the animistic child. The mystic can describe energy sensations in the body, the streaming and excitation of the orgonotic field, he may even give details that are astonishingly correct. But he will never quantitatively grasp these sensations, in as little as you can put the mirrored image of a log on a balance. (Id., pp. 96-97)
>
> Clinical research has shown that mystics suffer from a built-in wall between organ sensation and the objective process of stimulation [in their organism]. This wall is real. It is the muscular armor of the mystic. Any attempt to get the mystic in direct touch with the sensations of

the living triggers fear or loss of consciousness. The mystic can see the mirrored image of an emotion, but he cannot feel it as a real sensation. I state this so clearly here because my knowledge is based upon experience: if, namely, through orgone therapy, the armor is dissolved in a mystical thinker, all mystical experiences at once disappear. The existence of the separating wall between stimulus and sensation is the cause of the mystical experience. (Id., p. 97)

Mysticism vs. Spirituality

It may become clear through these elucidations that most people in our world are mystics and mystical thinkers, and especially those *hyper-rational reductionist thinkers* who deny emotions, and also those who deny children's emotions and sexual feelings. This is so because they are armored against their own natural body sensations. And this armor they have built is the embodied wall of their religious and ideological beliefs that serve them to repress their natural emotions, sensations and orgonotic streamings.

Once it is understood what mysticism really is, true knowledge-based spirituality can come about in the human mind. Carl-Gustav Jung, when asked if he was a religious thinker, or if he believed in salvation, simply replied 'I do not believe, I know.'

Spirituality is an extended realm of knowledge that reductionist science has not integrated so far. But this is currently changing. We are right now living through an era of strong developmental change that brings science and religion closer to each other by fostering and developing integrative views and publications that are the building blocks of a truly holistic science.

The following quote may demonstrate that Reich was not against true religion in the sense of *religio*, the connectedness to spiritual guidance, and that he did not hold that all religion was per se mysticism. Let us not forget that Reich wrote a study on Jesus Christ, a study that never was published and that he completed shortly before he died, entitled *The Murder of Christ*. Reich was not an atheist. He possessed that kind of intelligent and natural spirituality that we can find with one of his even more well-known scientific contemporaries, Albert Einstein.

> WILHELM REICH
> The widespread view that nature is basically harmony is essentially an animistic view that, however, the mystic

degrades into the idea of a personified godhead or the omnipresence of god. The mystic is as it were stuck in the absolute. The absolute cannot be grasped. The animist, by contrast, is a flexible thinker. He can change his views. And he also bears the advantage that his grasp of nature, contrary to the mystic worldview, contains a useful grain of truth. The animist Kepler, who discovered the planetary harmony and who explained it with the term vis animalis, was right, even from today's perspective, centuries after his death. The same energy that directs the movement of the animals and the growth of all living substance indeed also directs the stars and planets. In the functional identity of organismic and cosmic orgone, we find the origin of all animistic and truly religious worldviews. (Id., pp. 98-99)

Mysticism, Insanity, Cruelty, Brutality, Perversion and Fascism

Mysticism is not just a fancy, day-dreaming or *idealistic thinking*, as it is often called. It truly is a pathology, an aberration from nature, and a pitfall of perception.

Most politicians around the world are mystical thinkers; most people who are entrapped by sects, saviors, gurus and churches are mystics, and among scientists and even among hardcore rational computer programmers I have met many hidden mystics.

THE DESTRUCTION OF THE NATURAL ORDER

They all have in common that they are emotionally starved, incomplete, fragmented, in a constant intellectual overdrive, repressed, reductionist in their worldview, reducing namely all of life and living to some Darwinist evolutionary beliefs, often fond of genetics and affirmative of euthanasia, often *narcissistic* and most of the time icy and aloof in relationships. They are most of the time male and in relations seldom warm and empathetic. The female's organism is resisting much more than the male the specific *paranoid split* which is part of mysticism, and that is brought about through a fundamental split between ratio and emotions.

And there is more to it. Mystics also experience a basically distorted sexual life, be it perversion in the sense of a strongly repressed, ritual-based and sadistic sexuality, or plain impotence. Adolf Hitler is a famous historical example of an impotent mystic whose brutality and criminality is known today as a matter of popular knowledge.

The crux is that in popular wisdom, people like Hitler and so many of his Nazi followers are considered as rational-minded 'solution-givers' because the mass mind functions exactly on the same mystic wavelength as those perverse and abject political leaders. And here we face the tragedy of all human history and political history. Wilhelm Reich explains why this is so:

Wilhelm Reich

The mystical experience is rarely to be found without brutal-sadistic impulses. Further, in my experience orgasmic potency is seldom to be found with mystics, and mystical thinking is not to be found with orgastically potent individuals. (Id., p. 97)

Mysticism is a blockage of direct organ sensation and a projection of those sensations upon 'supernatural forces.' This is true for the spiritualist, the schizophrenic, the religious physicist and for any form of paranoia. (Id., pp. 97-98)

A good part of the brutality of the mystic can be seen in the fact that while he senses the living in himself, he does not really experience it, and therefore cannot let it unfold. He therefore develops a violent impulse to conquer what he sees like a spooky image in the mirror of his distorted mind, so as to make that faint shadow touchable and real. For the mystic mind, the mirror of reality is a steady provocation that drives him into a frenzy. Here it is, the living, the moving, the laughing, crying, hating, loving emotional being … but only in the mirror, in reality unreachable for the ego of the mystic, as the fruits, in the old myth, were for the tortured Tantalus. From this tragic situation arises every impulse for murder that is directed against the living. (Id., p. 122)

THE DESTRUCTION OF THE NATURAL ORDER

Narcissism

I have learnt about *narcissism* at first through the books of Alice Miller and Alexander Lowen, back in the 1980s. Both psychiatrists were since long years specialized on narcissism and it was through their unique input and unwavering efforts that today the seriousness of the narcissistic affliction has been recognized in mainstream psychiatry.

This was namely not the case when they started out to publish on this matter, back in the 1970s. At that time, narcissism was as good as overlooked in psychiatry, and was not thought to be a serious affliction. Today, while health care professionals recognize the seriousness of narcissism as a psychiatric disorder, the general public maintains a state of confusion and misinformation about the very term and the nature of the narcissistic affliction that I have hardly seen for any other psychiatric problem.

It's dumbfounding when you see that popular encyclopedias explain narcissism with assumptions that actually are pure nonsense because if narcissism meant abundant love of oneself, there would not be a problem. But fact is that narcissism is the very contrary of love of oneself, it is the very denial of love of oneself—and that makes that it's a problem.

What is Narcissism?

Perhaps it was an advantage that I never bothered too much about the term itself, as it is confusing and misleads many people.

There is about no other subject where the clash between professional knowledge and the half-knowledge of lay persons is so large as with narcissism. Everybody seems to know what narcissism means, but when you inquire further, you see that people maintain the strangest misconceptions about this pathology.

Most people have heard about the ancient myth of Narcissus that is at the origin of the term. But what does this myth tell us?

Here is where the misconceptions start. Most people somehow got an idea and extrapolate from the little knowledge they have, and the result is a standard answer like:
—Oh yes, this strange guy who looked in the water and saw his mirror! That guy loved himself too much,

THE DESTRUCTION OF THE NATURAL ORDER

he was fallen in love with himself … and then they go concluding narcissism was a hangup of people who 'love themselves too much,' who are fixated upon their own self-image, who are fallen in love with themselves.

Needless to say that all of this is sheer nonsense. The very contrary is true. Narcissism is a pathology where the person, through a deep hurt suffered early in life, is unable to love himself or herself, and thus lacks even a basic level of self-love. And what is worse with this affliction is that the true self of the person, their self identity, their feeling ego, their *Me*, and also their body image, have been buried deep down in the unconscious. The result is that narcissistic people do not know who they are or, as it is expressed in psychiatry, they deny their true self.

This denial of their own intrinsic being, their character, their values and oddities, their depth and dignity is what lets them appear like shadow dancers. They are generally fluent talkers and take up new ideas quickly, but they don't integrate novelty, because there is nothing they could integrate it into, as they are *out of touch with their true identity*, the fertile soil of their human nature, their grounding.

NATURAL ORDER

I use to call them for this reason *narcissistic comedians*, as they actually behave as if being on stage, as if life was a huge stage where everybody performs a role – but where nobody plays the role of himself or herself, but always another. A plays B, B plays C, C plays A. But life normally is that A plays A, B plays B and C plays C.

People who suffer from narcissism tend to appear *aloof*, they appear to *float*, as if their feet never touched the ground. There is often also something *Peter Pan* like about them, something fragile and strangely youthful, often accompanied by a sunshine smile that seems to suggest that they know no sadness. While in truth, they are the saddest people on earth, only that they *can't even feel their sadness*, alienated as they are from their feelings.

In exchanges with narcissists I also found that they often deny the reality of emotions, trying to grasp all of reality with their pure intellect—that usually works brilliantly well. But that makes that they are truly alienated from humanity because they more or less consciously discard the *irrational* out of the world.

For narcissists, *all must be rational, clear and straight*, and they tend to condemn irrationality in

THE DESTRUCTION OF THE NATURAL ORDER

people, out of touch as they are with their own irrationality.

We humans are at times rational and at times irrational. We are as good as never only rational or only irrational; we are a steady mix of many qualities and oddities, and it's our vivid emotions that bring the necessary kaleidoscopic change in our lives so that we are not for too long rational and not for too long irrational. But for the narcissist there has to be only rationality, and all the rest is as it were human weakness ...

How To Identify Narcissism?

You can identify rather quickly if you suffer from a narcissistic fixation. Simply check if you play yourself in your life, or if you play a role that fakes it is you. Then, when you ask this question and it rings like 'But who is me?', you are getting on the right track. When that question feels odd and strange because *somehow you have never asked who you are*, and if in the game of life you as good as never play the Me-card, then you know you have a problem with narcissism.

Another reality check would be the *obsessional idea to be altruistic* and 'always good' to others, to a point of self-forgetfulness. If that rings true to you, you should check if it's a moral duty for you to be always concerned about others and to put yourself behind. If this is the case, you probably have a hangup with narcissism, and you are denying your true self, replacing the vacuum in yourself with the vacuum at need with person A, friend B or relative C that you have to help out, to save from bad luck, rape or incest, to heal, to comfort, to look after, to console, to protect, and so on.

Narcissism is really not a complicated pathology and it's not difficult to grasp. It has been made difficult to understand through popular psychology that loves to use strange and fancy terms and abhors to express simple things in a simple way. It's much more difficult to explain what neurosis is or psychosis than to explicate what narcissism means and what makes persons afflicted with narcissism suffer so much in life. *They really suffer!* Narcissism is not a party affliction, not a gentleman's ailment, and not an outflow of vanity, while it is often belittled as such. Narcissism is an affliction serious enough to be put on

THE DESTRUCTION OF THE NATURAL ORDER

priority by most of today's psychiatric services. For when you're out of touch with yourself and your deepest emotions, you live a life that is not yours, you live as it were an empty life. This inner vacuum, this emptiness when it's constant is something that can trigger other serious afflictions such as substance abuse, chain smoking, depression, chronic fatigue, alcoholism, anxiety, phobias, and sexual obsessions, aggression and perversion. It also can trigger somatizations, which means that the body gets ill for reasons that are not physiological, but merely psychological.

NARCISSISM AND SOUL

Another corner of the vast literature on narcissism is what spiritual-minded people say about it. Their terminology is different, and that unfortunately also contributes to the general confusion about narcissism.

I have in mind a particularly successful and brilliant author, Thomas Moore, whose most famous bestselling book, *Care of the Soul (1994)*, basically is a manual for healing narcissism.

But the problem is one of terminology. Moore speaks of 'soul' and of lacking soul when he describes narcissism. His ideas are brilliant, and he has pointed the finger on the wound when he says that narcissism cannot be healed through pushing the person into a growth cycle or by otherwise suggesting the person 'to grow up.'

> THOMAS MOORE
>
> Narcissism has no soul. In narcissism we take away the soul's substance, its weight and importance, and reduce it to an echo of our own thoughts. There is no such thing as the soul. We say. It is only the brain going through its electrical and chemical changes. Or it is only behavior. Or it is only memory and conditioning. In our social narcissism, we also dismiss the soul as irrelevant. We can prepare a city or national budget, but leave the needs of the soul untended. Narcissism will not give its power to anything as nymphlike as the soul.
>
> —Thomas Moore, Care of the Soul (1994), pp. 58-59.

I have coached narcissistic and highly problematic individuals over the Internet, free of charge, for a period of almost ten years, considering this as the 'social' part of my mission as a coach, and I found invariably that these individuals wait for society to accept them, instead of doing the first step and accept themselves. Moore explains:

> THOMAS MOORE
>
> What the narcissist does not understand is that the self-acceptance he craves can't be forced or

THE DESTRUCTION OF THE NATURAL ORDER

manufactured. It has to be discovered, in a place more introverted than the usual haunts of the narcissist. There has to be some inner questioning, and maybe even confusion. (Id., pp. 60-61)

And I made an astonishing discovery. I had myself a narcissism problem over many years, since my childhood actually, and it was not cured in a psychotherapy, but I could cure it subsequently, virtually by *talking to the trees*. It was when living in the Provence, France, I took the habit to go for nightly walks, and I would address speech to some of the trees in an alley with sycamores. There were three huge sycamores that I felt attracted to, and what I would do, late enough so that no cars would pass by, was to put my left hand firmly against the trunk of the tree, and talk to the tree, either by thinking or by whispering my ideas. Now, what happened to my surprise was that not only was I greatly energized through this unique kind of conversation, to a point to not being tired when coming home, but I also had dreams where the tree was talking back to me. And I learnt amazing depths of wisdom from these dreams. Now, I was of course very surprised when I found the following passage in *Care of the Soul (1994)*:

NATURAL ORDER

THOMAS MOORE

I suspect that this is a very concrete part of curing narcissism—talking to the trees. By engaging the so-called 'inanimate' world in dialogue, we are acknowledging its soul. Not all consciousness is human. That in itself is a narcissistic belief. (Id., p. 61)

And indeed, through my talking to the trees, I felt a sudden interest in shamanism and went on a spiritual quest that took me several years. I engaged in a tedious research on shamanism and went to Ecuador, two years later, in 2004, to drink the traditional sacred Ayahuasca brew.

—See Peter Fritz Walter, Consciousness and Shamanism: Cognitive Experiences in the Ayahuasca Trance and Theories of their Causation, Series Scholarly Articles, Vol. 4, 2015/2017, and Audible Audiobook 2017).

I left this initiation transformed. I have regained the whole range of magical beliefs I once fostered as a child, and this really has completely healed the narcissistic condition.

Now, Thomas Moore has put a particular stress in this book on the danger of collective narcissism and he investigates deeply in the culture of the United States of America, to identify it as a model narcissistic culture. Moore writes:

THE DESTRUCTION OF THE NATURAL ORDER

THOMAS MOORE

Nations, as well as individuals, can go through this initiation. America has a great longing to be the New World of opportunity and a moral beacon for the world. It longs to fulfill these narcissistic images of itself. At the same time it is painful to realize the distance between the reality and that image. America's narcissism is strong. It is paraded before the world. If we were to put the nation on the couch, we might discover that narcissism is its most obvious symptom. And yet that narcissism holds the promise that this all-important myth can find its way into life. In other words, America's narcissism is its refined puer spirit of genuine new vision. The trick is to find a way to that water of transformation where hard self-absorption turns into loving dialogue with the world.

—Thomas Moore, Care of the Soul (1994), p. 62.

When we look at how present-day America, with its strongly narcissistic government, faces this 'loving dialogue', we indeed see that the *puer spirit* is strong. In addition, Americans somehow like to choose their presidents among *puer personalities*, and that may one day result in a fatal outcome! Mature cultures choose mature leaders, senior personalities, people who have grown out from the cradle or from an adolescence where Peter Pan is the dominating archetype.

NATURAL ORDER

I have to think of Terence McKenna's views on the psychedelic revolution in *The Invisible Landscape (1993)*, and the need for looking over the fence, when reading in *Care of the Soul (1994)* that curing narcissism involves an expansion of boundaries:

> Narcissus becomes able to love himself only when he learns to love that self as an object. He now has a view of himself as someone else. This is not ego loving ego; this is ego loving the soul, loving a face the soul presents. We might say that the cure for narcissism is to move from love of self, which always has a hint of narcissism in it, to love of one's deep soul. Or, to put it another way, narcissism breaking up invites us to expand the boundaries of who we think we are. (Id., p. 63)

And here again, when we look at present-day reality in the United States, boundary-dissolving substances, from DMT over LSD to Marijuana have all been declared illegal, which shows the degree of narcissism at the top government level in the enlightened nation. Only that the light seems to come from the wrong source. Plus the enlightened nation is an action nation. All is action! Anthony Robbins, the major coach-actor of the nation performs in shorts, jumping around like a school boy. When all is action, everybody is an actor. Not himself. And everybody acts out his or her life, instead of living it.

THE DESTRUCTION OF THE NATURAL ORDER

This *timelessness* of the nation, which is embodied in its business values, business standing for busyness, is one of the symptoms of its cultural narcissism that is not a present-day phenomenon. The action-nation was born in New England. When there is no more time, there is no more soul. Moore explains:

> A neurotic narcissism won't allow the time needed to stop, reflect, and see the many emotions, memories, wishes, fantasies, desires, and fears that make up the materials of the soul. As a result, the narcissistic person becomes fixed on a single idea of who he is, and other possibilities are automatically rejected. (Id., p. 67)

Peter Pan resisted to grow up, yet astonishingly, Thomas Moore writes that growing-up is not a cure for narcissism, in the contrary:

> But the solution of narcissism is not growing up. On the contrary, the solution to narcissism is to give the myth as much realization as possible, to the point where a tiny bud appears indicating the flowering of personality through its narcissism. (...) Narcissism is a condition in which a person does not love himself. This failure in love comes through as its opposite because the person tries so hard to find self-acceptance. The complex reveals itself in the all-too-obvious effort and exaggeration. It's clear to all around that narcissism's love is shallow. We know instinctively that someone who talks about himself all the time must not have a very strong sense of self. To the individual caught up in this myth, the failure to find self-love is felt as a kind of masochism, and, whenever

masochism comes into play, a sadistic element is not far behind. The two attitudes are polar elements in a split power archetype. (Id., p. 71)

When we apply this truth to the Peter Pan nation, we learn that we have to let them run where they run and let them break even more glass everywhere in the world, right?

I am not sure if Thomas Moore wanted to say that because once of a sudden, after having expanded into collective narcissism, he again speaks of the individual. But our daily news about the hero culture really seem to suggest that Moore's analysis of collective narcissism, that is shared by number of depth psychologists, would lead to an abysmal accumulation of Peter Pan like acts, performed as a nation-narcissist on the world at large, in order to gain depth.

I doubt that this psychological solution is going to work out politically, because even the most optimistic of Peter Pans around in the great nation may get a hint of stretching the bow too much … and the international repercussions may not permit Peter Pan to continue his puer game infinitely …

THE DESTRUCTION OF THE NATURAL ORDER

Anyway, from the soul perspective, and leaving political realities untouched, Thomas Moore writes:

> The secret of healing narcissism is not to heal it at all, but to listen to it. (…) I am stuff. I am made up of things and qualities, and in loving these things I love myself. (Id., p. 73)

This is in accordance with a general soul-based healing approach that was the prevalent approach to healing during the Middle-Ages and the Renaissance. Moore writes:

> Robert Burton in his massive self-help book of the seventeenth century *The Anatomy of Melancholy*, says there is only one cure for the melancholic sickness of love: enter into it with abandon. Some authors today argue that romantic love is such an illusion that we need to distrust it and keep our wits about us so that we are not led astray. But warnings like this betray a distrust of the soul. (Id., p. 81)

The Origin of Narcissism

If we wish to realize our personal identity and become whole human beings, we have to be able, still in childhood, to form a genuine personal identity. This is however impossible if we are reared by narcissistic parents, those namely that are *indifferent to the unique person of the child* they have brought to life. But this is not all there is in the etiology of narcissism as a cultural perversion. We are formed and deformed not only by the influence of our milieu, our background, and our caretakers, but also by the influence of society as a whole.

Modern society is not set to let children grow into who they really are, but conditions children according to the requirements of a consumer culture, and molds them accordingly. This is primarily done through indoctrination and, secondly, through *gradually alienating children from their bodies*. The most effective way to indoctrinate children is to implant in their mind a deeply rooted doubt about *who they are*. This doubt which creates a vacuum will then be filled with magic formulas such as 'Be not what you are!'

The next step is to force the child to play roles for pleasing their parents. The main role in this drama which is the *Drama of the Gifted Child*, as Alice Miller called it in her book with the same title, is the role of the child as father or mother of their own parents. This education that I like to call 'rearing narcissistic comedians', is very common in *Oedipal Culture*. This is precisely why narcissism is rampant in Western nations, especially in the United States.

THE DESTRUCTION OF THE NATURAL ORDER

However, few researchers see that the main etiology of narcissism is to be found in our child-rearing paradigm. Those who do, such as Alice Miller or the late Alexander Lowen are not representing mainstream psychology, despite the brilliance of their work. They have, inter alia, found that education that typically leads to narcissism is rich in inventing and executing magic formulas that are given to the child within the context of 'good education' but that are in reality hypnotic injunctions.

These injunctions have been found by TA as highly destructive for the child's emotional, cognitive, motor, skill and sexual development. They are voiced often nonverbally, by implication, through examples given, through confused and imprecise language, through reproaches and through comparisons that may or not be true.

—Be adaptable and flexible until self-alienation;
—Never be yourself in front of your parents;
—Be not child-like, but adult-like;
—Be mature in immaturity;
—Understand what your parents don't understand;
—Be logical and uncomplicated;
—Respect your parents while disrespecting yourself;
—Mistrust your intuition;
—Follow authority without questioning.

I see another etiology of narcissism in the lacking symbiosis between mother and infant during the first 18 months after birth. Regularly, with mothers who themselves suffer from narcissism, clinical research found a *reduction or total absence of eye contact* between mother and child, absence of breastfeeding or when the breast is given, the mother feels revulsion, disgust or aggression toward the child; in addition, such mothers tend to be hostile to the child's first steps toward autonomy, thereby creating in the child a pathological clinging-behavior that has very nasty consequences later on in the development of the child and young adult.

Often what happens in such relationships is that the mother manipulates the child into a real *co-dependence* where she projects her longings for love that are unfulfilled in the partner relation, upon the child. This then in many cases leads to emotional abuse.

Narcissism thus is often the inevitable result of emotional abuse suffered in early childhood, and that fact may help to understand the gravity of the affliction of narcissism.

THE DESTRUCTION OF THE NATURAL ORDER

What this results in is that the person unconsciously later tries to heal the lacking primary fusion by repeated pseudo-symbiotic relationships, which are relationships where love is replaced by dependency or where love is confused with dependency. However, since those persons that are invested with that role of ersatz mothers and fathers can never give the lacking primary fusion, disappointment and depression will invariably loom over those relationships.

Narcissism is an inevitable by-product of patriarchy, and its etiology is *wrong relating.* Wrong relating to self. Wrong relating to others. It is built on what Joseph Campbell called the *solar worldview* and ignores the many shadows of the soul – and thereby ignores its own shadow.

Narcissists, therefore, are tragic figures. They are tragic in the sense that they run into the abyss *without the slightest idea of what they are doing* because they are not grounded and have their feet in the air, like the *Fool* of the Tarot. They are lunatics, because they have not integrated their own Luna, their Moon energy. They are the eternal Peter Pans of sunshine movies, and present themselves to the public smiling, broadly smiling, most of the time, but in haphazard moments you see their true face—while they themselves ignore it.

Denial of Complexity

The Etiology of Fascism

A typical characteristic in the destruction of the natural order, present in all fascist and totalitarian regimes, is a denial of complexity. This means in practice that natural complexity is replaced by overly simplistic reasoning for explaining facts of life, or certain sociopolitical realities. As Jacob Burckhardt, the Swiss historian, once put it:

—The essence of tyranny is the denial of complexity.

Complexity and Simplicity

Complexity is a major characteristics of all living systems. In all flow patterns, *complexity* and *simplicity* are complementary opposites. This is so not only in natural phenomena, but also in ontology and in human psychology.

THE DESTRUCTION OF THE NATURAL ORDER

This duality that is inherent in human psychology has very well been recognized by the ancient Mesoamerican natives. As Mary Miller and Karl Taube write in *An Illustrated Dictionary of The Gods and Symbols of Ancient Mexico and the Maya (1993)*, under the header of 'duality,' one of the basic structural principles of Mesoamerican religious thought is the use of paired oppositions. They write:

—Mary Miller, Karl Taube, An Illustrated Dictionary of the Gods and Symbols of Ancient Mexico and the Maya, London: Thames & Hudson, 1997.

MILLER & TAUBE

In these pairings, there is a recognition of the essential interdependence of opposites. This complementary opposition is most clearly represented in the sexual pairing of male and female. To the Aztecs, the supreme creative principle was Ometeotl, the god of duality. In this single self-generating being, the male and female principles were joined. The omnipotent god could also be referred to by its male and female aspects, Ometecuhtli and Omecihuatl. Similarly, the Mixtecs and other Mesoamerican cultures considered creation to be the work of a sexually paired couple. Aside from the male and female principles, common oppositional pairings include life and death, sky and earth, zenith and nadir, day and night, sun and moon, fire and water. It can readily be seen that such series of pairings could be easily linked into a larger group of oppositions. Thus, for example, one side could entail male, life, sky, zenith, day, sun, and fire, whereas the other side would be female, death, earth, nadir, night, moon, and water. Such larger structural oppositions are evident in both contact period and contemporary Mesoamerican religious systems. (Id., p. 81)

Complexity and Consciousness

Further, it is important to realize that there is an intricate relationship between *consciousness and complexity*. Villoldo and Krippner note in their captivating study *Healing States (1984)*:

> VILLOLDO/KRIPPNER
>
> The term *consciousness* is used to refer to a person's overall pattern of perception, thinking, and feeling. Some specific patterns, or *states of consciousness* appear to be especially conducive for self-healing or to the healing of others. These healing states require scrutiny whether they involve mediumship, shamanic phenomena, or any other experience that is of potential value. (Id., p. 21)

Structure and content that I assign to consciousness basically consists of three major elements:

- Perception
- Information Processing
- Energy

THE DESTRUCTION OF THE NATURAL ORDER

The most important part of my scientific assessment of consciousness is that it contains the *zero-point field* or *quantum vacuum,* so that the human energy field is a constituent part of it, next to perception and information processing.

In Western scientific history, the fact that consciousness is a total information field has been blinded out from scrutiny and occulted, to a point that in Western cultures, there is a huge knowledge gap about the cosmic energy field as a result of this cultural prohibition of the 'tree of knowledge.'

Consequently, my consciousness research is focused upon bringing in the missing links again so as to arrive at a unified field of integrative perception and thus a coherent model of consciousness.

It is evident that the most important part of consciousness is not perception or information processing, but the *cosmic energy field*. Hence, complexity is a function of the energy flow; when energy flows freely, complexity tends to be high, while it's reduced when energy is blocked or obstructed. (See graphic illustration on the next page)

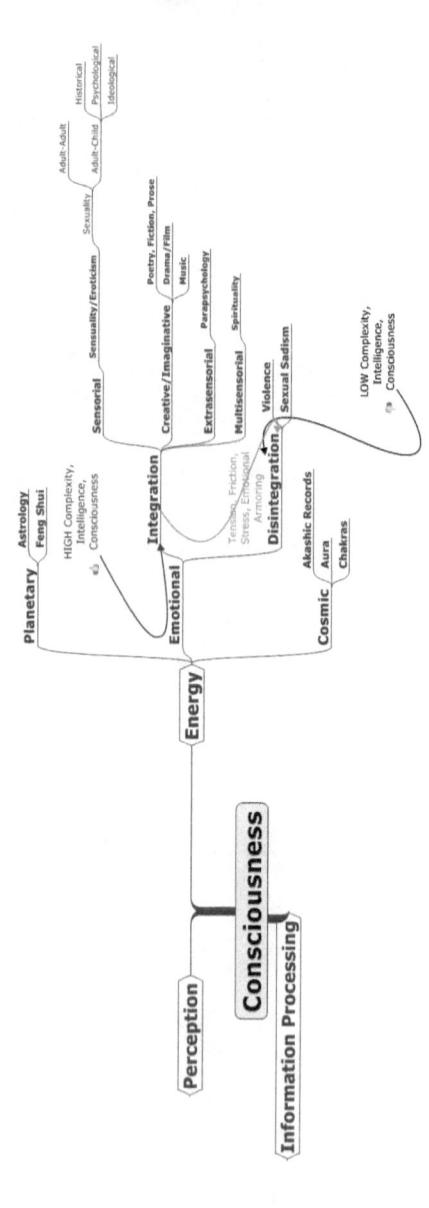

As a matter of evolution, life and particularly human life *tends to increase in complexity over time*. However, historically and socially, every time when shifts of consciousness occur that bring about a marked increase in complexity, a counter-reaction sets in that typically, and propagandistically, *denies complexity* and begins to threaten, persecute and socially discard individuals, and especially scientists who research in areas of human complexity that are not yet fully understood and that are therefore surrounded with taboo, confusion and fear.

Social historians such as Jacob Burckhardt, and psychologists have found the deeper reasons of fascism in a deep-rooted fear of change, and an almost paranoid fright in facing complexity, and especially *erotic complexity*.

Complexity and Child Abuse

We find in child abuse generally a denial of complexity, as the child for the abuser is an object for self-gratification and in addition a provider of emotional security, and thus a thing to manipulate; behind protection is often hidden a deep longing for domination and manipulation.

NATURAL ORDER

When the child is seen as a complex being, where the difference to adulthood is merely one of learning certain skills, and when the child is recognized as being a *complete human* emotionally and sexually, child abuse will decrease! And logically so, the contrary is true as well: where a society denies the child this complexity by declaring it a 'nonsexual' creature, or a being that has no sensual longings of its own and a brute body that can be ruthlessly beaten upon without 'really feeling the pain,' then you automatically raise the incidence of emotional, physical and sexual child abuse.

The latter paradigm is the one today practiced by international consumer culture that vehemently, through all its media, its governments and its NGOs requires to protect children from abuse. However, what this culture does is exactly the contrary in that all their government and non-governmental policies for *child protection* do only one thing for sure: they *raise child abuse*, by raising the probability for child abuse to happen, because they *need child abuse to happen.*

Their protector industries are businesses and run as businesses, and they have to have a cause tomorrow, otherwise they will run out of funding. Thus more child abuse tomorrow ensures them to be successful, and not less child abuse tomorrow.

—See also Peter Fritz Walter, The Commercial Exploitation of Abuse: A Study on Policy, Essays on Law, Policy and Psychiatry, Vol. 10 (2018).

The Denial of Erotic Complexity

Under *erotic complexity*, I treat under one unifying header three different forms of complexity: *affectionate, emotional* and *sexual* complexity. As the conclusions regarding their complexity patterns is pretty much alike, I unify the three paragraphs into one.

Let me first of all clarify what affections are. I treat in my writings affections and emotions in a synonymous way while I am conscious that in psychology there is a trend to distinguish between affection and emotion. Sexuality, for the vulgar and ignorant masses, is an animal-like behavior they do like automatons while never talking about it. By contrast, for the spiritually awakened and conscious individual, sexuality is the highest form of religion, of true union with god, the god in us and the god in our loving partner. Thus, sexual copulation is a fusion of god with goddess, and can bring about, when used correctly, the highest form of enlightenment for both partners.

In the ancient Indian tradition of sexual *Tantra*, this wisdom has been orally transmitted from generation to generation but is today dried out as *Vedanta* is by itself a perversion of the original *Tantra* religion of India in that it denies in origo the sexual nature of religion, replacing it by an exaggerated and hypocrite moralism that is based on so-called devotion or Bhakti, which is but an artificial and powerless surrogate for the original sexual copulation with the god or goddess as it is still part of native religions such as Huna.

But Tantra is misunderstood in the West. It has originally nothing to do with sexuality in the vulgar sense of the world, nor has it in any way a connotation with immorality, in the sense of licentious or even criminal behavior. In the contrary, it is in the Roman sense a *school of virtue* and teaches the right usage of sexuality as a god-given and god-driven force that unites us with ourselves, others and the cosmos, and ultimately with the creator force. In this sense, tantric sexuality is holy, is religious, and is sacred!

THE DESTRUCTION OF THE NATURAL ORDER

What happens namely through the tearing down of the sacred sexual force into vulgarity, shame and moral corruption is that the resulting 'sex' is used for debauchery, for corrupting others, for example children, for corrupting a whole community of people when used for entertainment, pornographic parties and group sex orgies that are devoid of any link to the soul, and therefore contrary to religion.

This is how sexuality as such became corrupted in the process of the destruction of the natural order. The patriarchal mindset, religions and lifestyle is the main culprit here, and has ultimately raped and ravished the goddess. This is the collective guilt and shame accumulated by dominator culture that today has become a 'fuck culture' sharing in the group-fantasy of orgiastic swines who engage in swinish behavior when 'doing sex.' Pigs who think and stink of pigs, that's what our society is, when it addresses the most holy behavior of all human behavior, the sexual function, copulation, unity.

However, as our current dominator societies, since about the beginning of patriarchy five thousand years ago, are ruled by ignorant and vulgar politicians, the masses, instead of being enlightened by sexual union, are more and more perverted into a shame-based identity mainly produced by sexual guilt, and their souls are abducted by organized political manipulation into truly devilish behavior, repressing the best in them, their true talents and gifts, together with their holy sexual drive that they shun and curse like a venom in their mind and blood.

All major dominator religions have perverted the very root of man, the original innocence of humans, and have by so doing dried out the fertile source of growth which after all is sexuality and sexual strength, together with emotional integrity and spontaneously virtuous behavior toward others.

They have contributed to the emotional starving of modern man and prepared the ground for rampant sexual impotence, emotional and sexual frigidity and true debility of the consumerist masses worldwide; they have prepared the soil for the most horrible form of cancer there is, *emotional cancer*. The most popular form of this cancer, that Wilhelm Reich called the *emotional plague* is our modern child protection paradigm.

The truth about our sexual origins and emonic integrity is veiled under such mountains of crap and

THE DESTRUCTION OF THE NATURAL ORDER

lies that it is almost impossible to communicate this truth today to any living human, except highly developed sages, on one hand, and witches and sorcerers, on the other—who are after all equally highly developed beings, only that they for the most part work on the dark side of creation.

THE DENIAL OF CHILDREN'S EROTIC COMPLEXITY

Children in today's international consumer culture are emotionally manipulated and their souls abducted from their most tender age, and this in the name of 'good' education, morality and so-called *child protection*, and of course always 'for the best of the child.' While it's in truth for the best of nobody, as it brings only perversion, domestic, collective and structural violence, shame, dishonesty and rampant individual and collective schizophrenia.

Originally perennial science was not separate from religion which is the reason why strong and redundant native religions such as *Huna* are truly scientific. They have understood and taught the innate complexity of emotions and sexuality, and generally of all relationships between living beings. This is why they have never discarded out from their scientific worldview extrasensorial perception, telepathy, prophecy, precognitive dreams and spiritual healing.

Modern Western science developed a *highly dysfunctional sub-science* under its general header of psychology that calls itself arrogantly *sexology* but which is a mechanistic robot science that considers the human being as a sexual automaton, discarding out the emotional dimension in sexual attraction. But even this reductionist and highly Cartesian science eventually could not hide the truth that human sexual behavior is immensely complex and unpredictable for the most part.

So when you see the larger picture here, a true *emosexuality*, a sexuality that is correctly linked and backed up by integral, correctly vented and functional emotions, then you will get a hint of how complex the human being is, and how stupid and reductionist most of today's psychologists, social policy makers and politicians are.

The Plague of Sadism

The Etiology of Sadism

Sadism is a blockage of the natural *emotional flow* through a predominantly moralistic or puritanical education, often accompanied by physical punishment, which leads to a *repression* of the natural streaming of the hot and melting sexual energy and as a result, to demonic emotions, and violence, because the *naturally deep sexual discharge becomes shallow* or even is inhibited. As a result, the naturally hot and tender sexual feelings are disintegrated and distorted into a *compulsion for sex* targeting at strong explosive sexual discharge, as a matter of abreacting an urge, instead of embracing a mate. The sexual discharge temporarily alleviates the fear armor but tends to entangle the person, who is unconscious of the affliction, long-term in sexual aggression, assault and a bullying, racketing or abasing behavior, that degrades and dehumanizes the mate, relegating him or her to a passive dummy.

Sadism was badly understood before Wilhelm Reich's in-depth research on the sexual orgasm revealed that the natural sexual response is by no means aggressive or compulsive, but controlled by empathy and love for the sexual mate. Wilhelm Reich states:

> Sexual responsibility is automatically present in a healthy, satisfying sexual life.
>
> —Wilhelm Reich, Children of the Future (1984), p. 208.

Only in sadism, which is a distortion of the natural emotional and sexual setup, this empathy tends to be overridden by an overwhelming longing for egocentric, and power-ridden satisfaction virtually on the back, and to the detriment, of the sexual mate.

> It is a fact that only the person who is incapable of gratification, the person whose sexual life is impeded and disturbed and who is contaminated by moral inhibitions, becomes sexually dangerous, while the sexually gratified and healthy person, no matter how many and what relationships he has, poses no risk to social coexistence. (Id., p. 193)

The Abuse Pattern

This is why long-term sexual sadism leads to a corruption of the personality, as the pattern for abuse then is laid also in a general manner, and the person tends to take advantage of others in the form of a habitual behavior structure, and thus becomes what is

THE DESTRUCTION OF THE NATURAL ORDER

called an 'abuser.' But for this to happen, the pattern must have been ingrained for long, and the person must never have gained awareness about it. This is rather the extreme case, as often people become conscious of their sadistic needs and begin to become suspicious about the obvious violence of their sexual behavior, and then begin to look for a way out, and may seek out a minister, physician, psychiatrist or psychotherapist for advice and consultation.

Breaking the sadism response is facilitated by being around babies and small children, and generally, when men are actively involved in taking care of children, of trees, of gardens and flowers, or for cooking and cleaning the house. Hence, the need for involving males in early child care. All these tasks are getting men in touch with their *yin* side, or *anima*, thereby helping them to overcome the macho or hero spirit that is negatively conducive to building the abuse pattern as a long-term affliction and personality trait. For we have to see that sadism is not only an individual problem, but also a societal concern. As early as in 1949, Wilhelm Reich wrote in his book *Ether, God and Devil*:

Natural Order

Wilhelm Reich

The unarmored organism does not know an impulse to rape and murder little girls, or to get pleasure through violence. It is therefore indifferent toward all moral rules that try to repress such impulses. It cannot comprehend that one has intercourse with another only because there is an opportunity for it, for example being in one and the same room with a person of the other sex. The armored character, by contrast, cannot envision an orderly life without strict moralistic rules against rape and lust murder.

—Wilhelm Reich, Äther, Gott und Teufel (1983), p. 76 (Translation mine).

If our Western culture was not largely sadistic, we wouldn't face the sad reality that virtually every day, in one of our glorious nations, a little boy or girl is abducted, raped and killed, or disappears under mysterious circumstances. This sadism can be shown and demonstrated also with many examples from the historian's or the psychohistorian's toolbox.

Sadism and Moralism

Sadism is a direct outflow and consequence of centuries if not millennia of moralism as a sort of emotional plague that has distorted our emosexual behavior structure. Our value system is deeply freedom and touch-hostile and this value system was

THE DESTRUCTION OF THE NATURAL ORDER

built because our deep emotions are out of touch with our natural emosexual base structure.

This value system is against nature because it favors violence and shuns natural sexual tenderness and respectful nonviolent embrace among generations, as a prolongation of necessary and health-fostering touch among all members of society.

Conspiracy Thinking vs. Critical Thinking

Conspiracy Theory is a hypothesis that alleges a coordinated group is, or was, secretly working to commit illegal or wrongful actions, including the attempt to hide the existence of the group and its activities. This is the dictionary definition.

What I am saying is that the natural order knows the self-thinker, while the destruction of the natural order brought us the uncritical nerd who is halfway paranoid, and lives in isolation and despair, avidly out for conspiracies to happen, so that he or she can feel involved in the political discourse. Needless to add that this kind of involvement is false, as it is based largely on myths, on hearsay information, if not on lies and slander.

Generalities

One of the main targets of conspiracy thinking are *secret societies*, especially international franc masonry. Before discussing the hairy topic of conspiracy thinking, let me say I do agree that secret societies are a queer element in a democratic society. However let's look at the roots first.

Secret societies came up a few hundreds of years ago as a philosophical movement in times where there was not much democracy around, but rather, a certain danger for philosophers to say certain things about the rulers. Mozart was a free mason, Goethe was one, Emerson was one, and many other people we would judge as benefactors of humanity. I have had a very good friend in Switzerland, a Scottish who was a freemason and who taught me much about it. He was the most caring human I have seen in my life, was doing much good and had very balanced opinions, the contrary of a fanatic, unbalanced, chaotic or secret person. Theosophy was also a kind of secret society. Why? It was very dangerous at that time to attack organized religion, not as today where those religions are fading off into nothingness and are but a joke for our young people.

Not so in the 1920s. Somebody interested in an overarching spiritual truth, studying *comparative religion* was being judged a pagan or heretic, worse a dangerous element in society, almost a terrorist as we would say today. Hence, the need to find support and discussion in secret societies.

The other aspect was the old *Hermetic Tradition* that was hermetic, as the name says, in ancient times, and for good reasons. Not all truth is for all people. And that tradition that I studied extensively, has really benefited humanity, and much wisdom is contained in it.

Dangers of Conspiracy Thinking

While American society professes to be transparent and liberal, there is nothing farther from the truth. So what the heck have these people to rumble about secret societies? I do not see their concern. A democracy also sets in place a certain freedom for groupings, only that today you would call it a Yahoo Group and formerly, because of historical reasons, it was called a secret society.

I see a certain danger in conspiracy thinking. Adolf Hitler was a conspiracy thinker, nothing but that, he couldn't see reality, not only because he had an IQ of 80, thus bordering debility, as we know today, but because he was through and through a mythic thinker, seeing secret societies virtually everywhere.

To replace a critical mind with a conspiracy mind is exactly what Hitler did, taking the *Swastika*, a Hindu religious symbol and turning it around, thereby making a devilish confession out of it. That is perversion grand style and so is, in my view, conspiracy thinking.

People, young or old, who lead a balanced and happy love life won't get on those paths. It all shows that the sex repression of the youth, as it's ruthlessly practiced in today's postmodern consumer culture, is and will be the cradle to all possible perversions to come.

THE DESTRUCTION OF THE NATURAL ORDER

THE BIGGEST SECRET

I am going to critically review a book by David Icke, *The Biggest Secret: The Book that will Change the World* (1999).

Pedophiles, Pedophilia

Icke uses the standard label *pedophiles*, without defining what he understands under this notion, and accuses certain politicians to be *pedophiles*. Apart from the fact that this is simply non-scientific as a person is not a label but may engage in activities that bear the label (confusion between activity and actor, an old trick in demagogic journalism), he perhaps voluntarily confuses *child rape* with pedophilia, while research on sexual paraphilias shows that child rape and pedophilia are characterized by *two different etiologies,* and are two entirely different sorts of sexual behavior. He is explicit enough to charge certain well-known politicians and leaders with a number of violent attacks upon a child or children, in the way that such behavior must be judged, from the explicit description, as blunt child rape. But he labels such behavior as 'pedophilia' and the men in question as pedophiles. This is simply unscientific and he uses here the technique of *defamation* he says is the major technique of those he exposes in his book. He is no better after all.

The Reptile Theory

David Icke has probably never heard of *mythology* and of the basics of *psychoanalysis.* He simply takes *metaphorical content as real,* which is really man-in-the-street-thinking *par excellence.* Icke seems to have no idea of the imaginal realm which is the very basis for mythology, and also the basis for religious thinking. It goes without saying that there is in all native cultures reference being taken to snakes and reptiles as metaphors for divine powers, as divinities, which is not to be taken as alluding to real life animals. Hence the psychological notion of the *mythic animal,* that probably escapes Icke completely.

This is all I can say, because when you see that his point of departure is actually based upon ignorance of psychological facts, and thus flawed, his whole rhetoric and proof of the existence of reptiles is pure nonsense. But to reach the mass mind, this was probably the strongest element in his rhetoric because it may be what most accesses the primitive mind of the masses, who are living in the same world of ignorance of psychological realities as the author.

THE DESTRUCTION OF THE NATURAL ORDER

Psychiatrist Robert M. Stein calls for this reason Christian thinking *primitive* because *it denies the existence of the imaginal realm,* or while admitting its existence, claims to control this realm by dogma and supervision of the believer.

> ROBERT M. STEIN
>
> Creative psychological development, individuation, is dependent on spiritual freedom. When we say, for example, a man has a free spirit, do we mean that he freely or necessarily transgresses the imposed manners, mores and taboos of his culture? I think not. But it does mean the freedom to do anything or go any place he desires in the imaginal realm. He is a man who has clearly distinguished the sacral, timeless world from the secular, historical world. He knows he can move with unashamed dignity among the gods and demons which belong to the mundane world. Such freedom cannot occur with a primitive form of consciousness in which inner and outer reality are governed by the same laws and values. In this sense, our Judeo-Christian tradition is primitive in that our thoughts and desires are subject to the same dogma, the same regulation, as our deeds. Spiritual freedom requires a break with biblical tradition and the development of a new form of consciousness – a consciousness which promotes the cultivation of imaginal freedom.
>
> —Robert M. Stein, Redeeming The Inner Child In Marriage and Therapy, in: Reclaiming the Inner Child, ed. by Jeremiah Abrams (1990), 261, 265.

NATURAL ORDER

I know this from many personal examples, especially from Germany, people I have known and whose thinking is rather right-wing and fascist. They reject psychoanalysis, they reject psychology and they argue like Icke. In their worldview, great scientists and discoverers like Freud or Reich simply are *swines*. Needless to add that they hate Jews, and are arrogant enough to voice that even in public. And in Germany such thing is possible, which is one of the reasons I left that country forever.

The World is Being Dominated by Five Families

It is of no importance who dominates the world, if that should be true; it is only of importance, supposed it is true, that the overwhelming majority of humans develop some kind of *critical sense*, not allowing those who dominate to do their business unhampered.

I'd say if it is true, we should first of all applaud them because they must really be smart if they have got there. Then, after applauding them, we can think about a future world where this can't happen again. By just shouting *Revolution, Revolution*, nothing will be changed, nothing will be achieved, except blood shed. If it is true, then something must be wrong with

THE DESTRUCTION OF THE NATURAL ORDER

our perception of reality, or let's say *with how we were taught to perceive the world,* and to fit in this world. We might namely think of establishing a quite unpopular and different form of *child protection,* namely a lifestyle paradigm that smoothly guides our children back to nature, to inner silence, so that they're not eaten by our manipulatory media, video games, and the hidden polemics and lies that are written in our school books.

Thus the true concern, as I show it with my publications, is *education* or a new form of education, or self-education, not polemics against those in power. Those in power will die and new ones will be in power. It's of no importance who they are as long as the majority of humans allows this to happen, focusing on large-scale entertainment and on body building instead of mind building, running for the newest junk and the latest gimmick, without giving importance to living intelligent and meaningful lives, and become conscious of their soul values.

Blaming People or Institutions

The forth issue I would like to mention is that Icke blames certain people, or certain institutions, instead of seeing the psychological root causes of why certain things go wrong in the world. Humans are not perfect but when they live in something like a natural order, they more or less can reach excellence, and human imperfection will not have outright destructive consequences. When one reads Icke carefully, one finds about the same morality scheme as in Westerns and crime movies, the good guy (first of all himself …) and the bad guys, corrupted government officials, free masons, bankers and so on. It's nice for the movie world to stick to this scheme, it makes cinema easy, it makes it easy to sell films to the masses of people who wouldn't like to learn about the complexity of life, and of human beings.

But when a man like Icke, who comes over, after all, as an intelligent and aware human, comes up with this scheme, then I can't help but think he may think business. What is his purpose? Is it perhaps just to make more money with 'selling more conspiracy to more people, at higher prices,' to paraphrase Coca Cola's Sergio Zyman?

Would it not be more honest to make readers understand the *psychological root causes* of our violent patriarchal setup and the consequences and the karmic boomerang connected to it?

THE DESTRUCTION OF THE NATURAL ORDER

Putting blame on people is not only lacking smart but it's misleading, because the power in the world is not really going out from people but from *concepts and ideas that human beings create and that are taken by the masses as eternal truth*. So the culprit of all evil is in fact language, that is the *misuse of language*, the use of language that is conditioned along the lines of ideologies, strategies, and political stratagems. People can be replaced, but a system of concepts, created by language, has a self-perpetuating force!

Icke is not only a demagogue, but he's simply not on track when he looks for the roots of evil. But he uses a simplistic vocabulary just as those he attacks, and that is why he is ultimately successful.

The masses do not understand the complex language of life, but the idiotic language of humans who are following idiotic concepts that they themselves don't understand. If a certain politician, for example, wanted to euthanize all people who love children sexually, and he does not use the hanger term *pedophilia* or define it for his cause, he will achieve strictly nothing. Truly, the evil or destructive effect on the mass psyche is effected by the *concept of pedophilia*, not by the fact that a certain politician, parliamentary or policeman dislikes childlovers.

Anti-Semitism

The fifth issue is that Icke's teaching is, honestly speaking, outspokenly anti-Semitic. In this respect he can shake hands with Adolf Hitler. And I am pretty sure that much of his popularity is due to this very anti-Semitism because the masses today in all consumer societies are fascist, most of the time in a hidden way, but they are.

Secret Societies

The sixth and last issue is that Icke has misunderstood the intention of secret societies. He blames secret societies to be the breeding lot of global fascism or the organizing network that brings together all the evildoers in the world. He doesn't mention that originally secret societies were founded as the very opposition to tyranny, the tyranny of patriarchy, namely, even before the Middle-Ages, but then they gained importance with the raising power of the Church at the time of the *Inquisition*.

When alchemy was considered a crime, secret societies were a necessity because otherwise those

THE DESTRUCTION OF THE NATURAL ORDER

early scientists (that's what alchemists were) could not have been in state to communicate across the borders of their village or clan, and their scientific knowledge would have been insulated.

Even in our days, the intention of most free masons is to do good, not evil, and even to oppose evil by a way of living that is in accordance with spiritual laws.

To follow Icke, Goethe, Emerson and Mozart would have been evildoers as they were free masons. His reasoning is not only childish, but it's *manipulatory* because he uses the simplification scheme of *homo normalis* to reject all that the 'peasant doesn't understand.' He uses, just like Hitler and many other tyrants and media manipulators, the strategy of *denying complexity*, that is based upon the ignorance of the masses for the sake of getting power over the masses, the power in this case, to buy his books, and to get many people subscribe to his conspiracy theories.

The secret societies were a considerable force of resistance during the *Nazi* regime and similar regimes in Europe in 1940 to 1945. I happen to have studied, during my 3rd cycle program, *European Integration*, the resistance movements against Hitler in Europe, and it is notorious that secret societies were a forum for their communication, which shows they are actually against totalitarian forms of government and the suppression of basic human rights.

I hereby do not deny that factually it may well be true that five single clans reign the world, but to repeat it, to counter this state of affairs, we have to *wake up to another reality* as the myths Icke puts up. We have to awaken to *nature*, and nature in us, in our bodies. I mean we have to a do away with moralism, with judgmentalism, with all activism targeting to judge and expose others, to blame certain folks, and to declare other folks as heroes. We have to wake up to being *authentic* and defend the values of *emotional and sexual freedom*, and children having the right to go through childhood without being emotionally, physically or sexually abused, and thereby *being able to building a soul identity*, and their unique soul power.

Then, as a growing group of mature and spiritually awake soul beings and individuals, we can prepare another course of destiny for humanity. But it's not possible by attacking those in power, for this is not only not effective, but truly destructive. The wise and virtuous know that their wisdom, and their integrity, will influence even the most evil leaders and tyrants, such as Petronius influenced Nero during most of his lifetime.

And Nero clearly murdered less when under the spell of Petronius' wisdom and poems!

Youth Fascism

Let me start with a *Youtube* example, one of thousands and thousands of videos in the same style. Note that this is a random sample. I needed less than ten seconds to find it, and it really pervades that community, it is all over the place, so to speak. When you take off the TV shows from that community, and the music videos, the art and dance, and the private videos, the core remains, and that core looks pretty much like the following sample, and includes the whole arena of conspiracy that is one big multifaceted and endlessly repeated fascist plot of mostly young people from the USA, Britain, Holland, France and Italy. Or let me put it that way. Those who live in a prospering modern country, with civil rights in place, and a functional government, but who are endlessly talking about their fascist government have a reason to do so. The reason is in my view that they secretly wish to be governed not by their present government, but by a *real fascist dictator*, because they secretly adore these kind of abusive leaders.

The reasons why this is so are to be found in my books. It's a sadistic fixation that turned into self-destructive and fatal masochism with many young people because they feel they are not loved, not given enough attention, and not promoted for what they really want to do. It's really, as history shows, a dangerous situation for any government, for this mob of youngsters may one day explode, unite, and then represent a major force of upheaval and social unrest.

First Example

User Name: AnonymousTruther

Title of Series: Jesuit World Financial Control 1/5

Quote from the audio track (3:22 - 4:30)

AnonymousTruther
Gentiles rule it all, and specifically white gentiles because I believe in the supremacy of the white race as

evidenced by the cultures that it builds for good, and evidenced by the horrible terrible conspiracies that it carries out for evil.

That's why I call this the pope's international white power structure centered in Rome; and thus the devil has chosen the white race for its greatest accomplishments, even as the Lord has chosen the white race for his greatest accomplishments.

White supremacy is a historical fact that we must accept if we are going to understand the present day. There is no such thing as universal equality among the races and nations. It doesn't exist in humanity, it doesn't exist among equestrians, it doesn't exist among nature; there is no such thing as universal equality anywhere, anytime, anyplace.

Needless to add that there is at that point no factual information at all, no evidence presented, no proof of actual wrongdoing by either the order of the Jesuits or the Vatican or any other alleged or implied persons or organizations. What there well is present is *racial prejudice*, very clearly expressed and voiced.

Second Example

President Obama

In the same style, and depicting the same racist values, there is a series of productions on *President*

NATURAL ORDER

Barack Obama, when he was just in the office for two months. These productions left me speechless.

Third Example

Encrypted Fascism

I have observed this phenomenon since about the change to the 21st century. What I have observed are the following details:

- Young people, not only girls but boys as well, are longer virgins than ever before in human history, lacking out on the very basics of loving intercourse and sexual activity, which is a direct result of society's fascist attitudes regarding child and adolescent sexuality and it's deliberate confusion with pedophilia and child abduction and rape.

- Only to search Google with the keyword 'child sexuality' throws me out 81,300,000 sites, most of them about child abuse, and you can be lucky to find a single site that really deals with the biological and psychological facts of child sexuality. I found one, but at the bottom of the page there is a Google Ad, and guess what it links to? Child Sexual Abuse. It must be so in a society that focuses on abuse instead of focusing on love, and love making, which shows how upside-down and alienated from nature modern consumer culture really is.

THE DESTRUCTION OF THE NATURAL ORDER

- Young people develop fascist attitudes and a basically fascist worldview without being conscious of it in most cases. This fascism is thus latent, and it is widely, in my view, the result of prolonged sexual virginity, and a negative attitude regarding sexual activity as such, which in this case doesn't necessarily come from a religious education. In my view it is the result of a total hypertrophy of the left brain hemisphere, that leads to an early intellectual overdrive, and a tendency toward neurosis, from early childhood. The home may even have been a liberal one, but the person may have developed a strongly negative view regarding sexuality or intimacy, or closeness with others. The main reason why this happens is that there is no incentive from home or school to seek out loving mates, so that the youngster encloses himself or herself in an almost incurable narcissism – which is a lack of self-love that goes pathological. Needless to add that, as I was showing it in my criticism of the Oedipus Complex, the fact that children are supposed to be auto-erotic only, and not to have active loving embraces (that is, intercourse), contributes to this isolation of our children and adolescents. Often, these youngsters are very good in school, over-performing almost all the time; they are the quintessential 'good boys' and 'good girls' to their mom, and they often excel in a sport or with playing an instrument; else they may handle the computer in brilliant ways, may be good programmers, or they are good in sales and marketing. These people are most of the time intelligent, but they are not erotically intelligent.

NATURAL ORDER

The problem is paradoxically that their intelligence, because of lack of a love and sex life, becomes in itself a problem, a pathology, and then develops a tendency toward evil, toward negative thinking and large-scale projections that serve to shield and defend the widely pathological lifestyle and worldview of those youngsters.

- This fascism is propagated largely under different headers than before, it is sold as neo-paganism, a nature-loving antidrug position, widespread conspiracy thinking with a fascist hateful regard upon secret societies, that is religious groupings, and large-scale techno-thinking, formulas like techno-shamanism, techno-fascism or even more strange-looking encoded or encrypted formulas that serve the same purpose as it served under a Goebbels, that is to denunciate and despise certain people, certain groups, certain minorities, certain countries or blocks of countries, or certain governments. This is done by putting ruthless labels upon those people, groups, countries, institutions or politicians.

- Another argument forwarded often by these youngsters is that a world government can only serve fascist purposes. They for the most part ignore the doctrine of national sovereignty as part of international law, and they also ignore that in Europe, hundreds of years ago, great philosophers such as Kant or Rousseau made peace plans that intelligently debated the evil effects of national sovereignty and therefore suggested states should make an end to war and get together to confer a part of their sovereignty upon a supranational or international authority, such as a world government. Needless to add that the European Union as well as the United Nations are fruits of these ideas, while we are not as far as having a world government yet. So I am wondering if those youngsters want to do away with the European Union and the United Nations as well? The question remains open and I am quite convinced the answer can be found, and I pretty much know in advance how it looks like.

- What confirms the above conclusions is that these masses of young people are widely anti-Semitic in their overall mindset, without of course voicing it, because that would defy their self-imagine as benefactors of humanity; many of them, it is true, are on the spiritual track and expect to make a large ascension during their lifetime, if they are not outright obsessed to be angels or aliens instead of what they most abhor, that is, to be simple humans.

- Through personal exchanges with a number of those people that are mostly to be found in countries like the United States, the Netherlands, Germany and France, I have become aware that what they most hate are people like myself, who are older and who have a good deal of direct life experience, who have had a vivid sex life as children and who therefore can tell them worlds of knowledge that they ignore. I have found through the hate and reject I received from these people that what they most hate are self-thinkers, people who are really independent of the system and who have developed their own worldview, and created their own reality, also, but not exclusively because they are financially independent. I have been rejected in the strangest ways by these people. They are violent through and through and society will one day wake up when the big explosion happens, for this human mud really is explosive!

- Their fascism is largely unconscious and blind in the sense that the young people who signup for those ideas are talking about themselves, to repeat it, as benefactors, not evildoers, as positive-minded and fully mature individuals, while they are in most cases not even able to have a single nonconflictual exchange with another person without throwing them their one hundred and one conspiracy theories over the head. I'd not be astonished to learn that some of them are still at thirty complete virgins, wearing pampers at night, and are sucking their thumb or otherwise need to be mothered for every little experience they are making. They are surely Oedipal Heroes and their narcissism is their pride and public smokescreen.

- This fascism also typically sees the world dominated by a single black-and-white wash such as 'The Banks,' 'The Jesuits' or 'Five Families,' or else 'The Rothschilds,' 'The Warburgs,' 'The Habsburg Clan,' 'The British Crown,' 'The Federal Reserve,' or certain bloodlines are blamed for all the evil in the world. To remind only that bloodline thinking is quintessential fascist thinking. Sane people think of human beings, not of blood, to remind only Hitler's notorious allusions to Blut und Boden (blood and earth), which was perhaps the forerunner of all those fascist ideas that ghost around in our young people today.

What these young people obviously never have learnt is psychology, mythology and symbolic as well as abstract thinking. When a journalist like David Icke tells them that there are reptiles around and famous politicians are shapeshifting into reptiles or dinosaurs in front of their eyes, they may believe it as one hundred percent true, instead of asking the question if those reptile and serpent figures, or obelisks and other mythic symbols seen on churches or public buildings are perhaps mythic animals, representations of a religious reality, of mythology, and thus *metaphors*? And then they might want to inquire in the past of humanity, and with today still existing native peoples and ask them what those mythic animals represent? The danger here is obvious that misguided youngsters may get strange ideas under the spell of those widespread conspiracies. What are

NATURAL ORDER

they going to do under the spell of those ideas? I think we should have a watchful eye on it!

CHAPTER THREE

The New Natural Order

The Eight Dynamic Patterns of Living

General

Several years of research on *shamanism* and aboriginal cultures provided the evidence for drafting the concept of *The Eight Dynamic Patterns of Living*. I found that these patterns are universally respected and applied by major tribal cultures all over the world, which is why native societies live in peace with all-that-is, respectfully and integrally, and why they are in harmony with nature and basically know no crime and major lifestyle diseases such as cancer, heart disease or immune deficiency syndrome.

NATURAL ORDER

It is because *these patterns are conscious with native cultures* that these peoples live peacefully, constructively and ecologically in accordance with their environment and the cosmos as a whole.

The evaluation of multi-disciplinary scientific research clearly shows that postmodern international consumer culture triggers worldwide destruction economically, socially, health-wise, military-wise, ecologically, and in other ways.

My research brought me to formulate the hypothesis that the true reason of this destruction comes from the fact that *all patriarchal dominator civilizations, without exception, have disregarded and even shunned every single of the eight dynamic patterns of living*. This is so because the *continuum* balance that the eight patterns provide simply is lacking in patriarchal civilizations; it is lacking in modern culture's philosophy, science, military policy, diplomacy and foreign policy.

From this insight it becomes evident that the eight patterns are ideally suited to be taken as a guide concept to be implemented in a more wistful international culture of the future, perhaps within a greater paradigm of *deep ecology* as Fritjof Capra and other ecologists have suggested it.

This would then have to be worked out on a joint-governmental and supranational level and as part of our presently evolving post-industrial global culture.

The Eight Patterns

1) The Autonomy Pattern

All peaceful tribal societies have in common that they give their children an *utmost of autonomy.* In dominator cultures, that today represent the bulk of large and typically industrialized societies worldwide, the lacking autonomy of the child is a *truly pathological phenomenon* that often takes the form of co-dependence, which I call *symbiotoholism* or *emotional incest* and in general the unhealthy fusionary clinging of members of the family, or as collective fusion through the identification with groups, organizations and ideologies. In fact, observing the growth processes in nature, we can see that autonomy is something built in all living, and as such takes part in all growth. For realizing the full array of our individual gifts and talents and become whole humans, we should build, still in childhood, a genuine personal identity. This is however impossible if we are reared by narcissistic parents, those namely that are *indifferent to the unique person of the child* they have brought to life.

2) The Ecstasy Pattern

All peaceful tribal societies have in common that they have a strong ecstasy pattern built in their lifestyle which makes them once in a while enjoy group events where the usual rules of conduct are more or less set aside. Usually, these events are characterized by magic rituals, the consumption of mind-altering entheogens such as *Ayahuasca*, and the partial or total disregard for the incest taboo or other sexual taboos. This principle was widespread even among major civilizations; still some decades ago, during the Carnival in Rio, it was not uncommon that sexual incest was practiced between parents and children. It is also quite probable that intergenerational sex, while practiced in very few aboriginal cultures, is being allowed on a larger scale also in less permissive cultures during ritual events that serve to liberate and cultivate individual and group ecstasy.

3) The Energy Pattern

Life is energy! This is recognized as a vital life pattern in all non-Western societies, and thus the greater part of the world. Oriental cultures were historically the most wistful in recognizing and applying energy patterns for healing, good fortune and positive relationships. The Chinese science of Feng Shui is perhaps the oldest distillation of this holistic knowledge into something we today call a *science* while traditionally Orientals tend to speak rather of *philosophy* or of *religion* when they talk about the perennial science of the bioenergy. However, even in the West, alternative scientists from Paracelsus to Wilhelm Reich have acknowledged the existence of the bioenergetic functionality not only of the human organism, but also the human body's relationship with the weather, the atmosphere and the cosmos as a whole. While in substance these researchers observed basically the same phenomena, the way they termed the cosmic life energy varied. Paracelsus spoke of *vis vitalis*, Swedenborg of *spirit energy*, Mesmer of *animal magnetism* and Reich of *orgone*. And since millennia this same energy was called *ch'i* by the Chinese, *ki* or *hado* by the Japanese, *prana* in ancient India and *mana* with the Kahunas from Hawaii and the Cherokee natives of North America.

Finally, parapsychologists universally agree that the motor of all psychic phenomena is to be found in our bioplasmatic and egg-shaped aura, an energy body of lesser density that we carry around our physical body and which could be seen as an extension of our bioplasmatic energy, as it is composed of the same bioenergetic charge that we find in the bioplasm. Emotions are streaming bioenergy currents that are a direct outflow of the cell's bioplasm. I call them *emonic currents*. They have their seat not in the brain, as modern psychology wrongly believes, but in the bioplasm and in the aura.

4) The Language Pattern

Psychoanalysis revealed us the development of *conscious and truthful language* as a primary condition to be met for the sublimation of our primary asocial instincts.

THE NEW NATURAL ORDER

Furthermore, peace researchers found that a lack of language and thus of communication is at the basis of all forms of violence, inner and outer. This insight has not only psychological but also political consequences. For it clearly indicates that only *free speech and democracy,* both within the family and the nation, insure maintaining peace and regulate our natural instincts and desires, so that they do not become asocial and violent through denial.

To everyone who says that we have democracy and yet are a violent society, I reply that we do *not have true democracy* and never had. Behold, violence only comes up when verbal communication is impaired! Shame is the single major reason why the communication about vital issues is blocked. When we feel ashamed about certain vital events in life, such as sexuality, we cannot freely communicate about these issues, because we are blocked or inhibited by the nagging feeling of shame that comes up every time we tackle the subject.

Lack of communication leads straight to violence; where the mouth is defended to talk, the body takes over the role of the mouth – and the fist talks! We all know this from history and from private experience, and yet there is very little general conscience in our society about the almost holy importance of dialogue, of communication, not only outside, in relationships with others, but first of all inside, in the relationship with ourselves.

Our large civilizations do very little to integrate the wisdom of language because they are *not conscious* about the power of the word. Tribal cultures, however, are wistful in this respect and generally dispose of an array of rituals that serve for exactly what in our civilizations we do within a psychotherapy, that is *putting words on things, events and feelings.*

5) *The Love Pattern*

All peaceful tribal societies have in common that they follow the *love pattern* and not, as most of the larger nations, the morality principle. The present state of violence within and between our larger civilizations, especially those with high morality is in my view the result of the despise of the love pattern and the widespread use, also and especially in politics, of *moralism*. With other words, it is the disregard of one of nature's highest principles, the principle of *biogenic self-regulation* that is at the root of global violence and the lack of love and true care among most of the peoples of the earth. It is the hypocrite manner of preaching peace and democracy by our false and opportunistic leaders while they treat natural love like the Biblical serpent. Wilhelm Reich, in his extensive research on the psychological roots of fascism found that it is the repression of our natural emotions, first of all by prohibiting our young generations the natural acting-out of their love desires that brought us at the border of the present abyss of fundamentalism, persecution, slaughter, genocide, war, civil war and worldwide terrorism.

6) The Pleasure Pattern

All peaceful tribal societies have in common that they acknowledge the pleasure pattern, for example in the way they educate their children. In planning the child's future, what counts is not the father's job, which is the typical dominator position, but the natural inclination, talent, motivation and interest of the child! By doing this, instead of projecting upon children their parents' wishes and desires, education could make sure that every generation does in most of their time what they really are gifted for.

The result would be both a high level of skill and motivation for profession and career. It is not astonishing that now also in modern nations the pleasure principle begins to be seen as the main motivating factor for a person's advancement in life. Suffices to read the biographies of great men and women to see that all achievement is a result of desire and *persistent acting upon desire* and that there is no better catalyzing agent than biological self-regulation that is based upon pleasure.

7) The Self-Regulation Pattern

All peaceful tribal societies have in common that they follow patterns of self-regulation or permissiveness in the education of their small children, and consequently restrain from inflicting violence in form of physical punishment upon them. The most peaceful of those tribal nations, the Trobriands of Papua New Guinea furthermore are completely permissive as to children's sex play and free mating games.

8) The Touch Pattern

All peaceful tribal societies have in common that they are conscious of the paramount importance of touch; that is why they care for maintaining free body touch among family members, nudity, and abundant

THE NEW NATURAL ORDER

tactile nutrition for infants and small children in the form of baby massage, baby-carrying, and naked co-sleeping with children.

In dominator cultures such as ours, the *ignorant influence of life-denying pediatricians* upon parents led to an almost complete turndown of parent-child caring touch. The results are devastating, to say the least.

Now we slowly begin to see the macabre results of the deprivation of tactile stimulation in infants as psychosomatic medicine reveals that our immunity against viruses depends on touch and that lacking touch, especially in childhood, leads to more or less acute immune deficiency and as a result to higher vulnerability for certain lifestyle diseases.

I go as far as saying that immune deficiency syndrome really is the result of wrong, ignorant and criminal pediatrics, a direct consequence of turning down caring touch bestowed upon children within the family, and through the ruthless *pedophilia persecution* also outside of the family. The result is virtually killing us! The stupidity of those psychological, legal and social policies is so obvious that I spare any further comment!

NATURAL ORDER

Cross-cultural research has clearly shown that early tactile deprivation is one of the major inducing factors for the plague of personal, domestic and structural violence in any given society. Considering these research results, to maintain an adult-child touch-taboo, within our without the family, is truly irresponsible. In this context, I maintain my alternative and perhaps controversial position that it doesn't matter if this touch is erotic or not, as long as it is caring and beneficial to the child, it's okay. Why making this difference at all? If touch is erotic, so much the better!

This is wanted by nature simply because most non-spiritual parents would not touch their children enough if they did not have erotic stimulation when doing so.

I bet our society needs to take a long and hot shower for getting here, for washing off the last remnants of the bastion of false 'christianity' as the ultimate recipe for perversion! Or, in other words, when we know that nature is right and religion is wrong, the choice is easy, at least for smart people, don't you think so?

THE NEW NATURAL ORDER

The Holistic Science Paradigm and Worldview

A Matter of Terminological Correctness

Holistic science is nothing new. It is truly the oldest of science traditions, was traditionally called *hermetic science*, thousands of years ago, and today is called *perennial science* in allusion to Aldous Huxley's excellent research on *perennial philosophy*.

—Aldous Huxley, The Perennial Philosophy (1970).

If this is a strange idea to the reader, which wouldn't astonish me as this knowledge is still today hermetic, I recommend reading Manly P. Hall's *The Secret Teachings of All Ages (1928/2003)*, which is about the best that was ever written on the subject. In addition, this precious and well-written book contains a wealth of references for further research.

I will demonstrate in this sub-chapter that there is no functional difference between Huxley's definition of *perennial philosophy* and the concept of *perennial science* as this difference came up because of a purely terminological confusion. To be true, in ancient times, *philos sophia*, the love of wisdom, to translate it literally, was considered the *Queen of Sciences*, something like an overarching or header science, while at that time, according to the prevailing holistic paradigm, was called *philosophy*.

Behold, that term was not defined in the sense it is used today, and done today, and was done already in the last three centuries, that is as a *system of intellectual speculation!*

To repeat it, initially philosophy was a science, that followed rigorous research principles, and had nothing to do with speculation. This was so until *Aristotle*, who actually was *the first philosopher who relied almost exclusively on intuition* to formulate his concepts, which is why I consider him as the first philosopher according to the modern definition; he was perhaps the first brilliant *speculative thinker* in human philosophical history. By the same token, I say Aristotle was not a scientist, not only in the modern definition, but also not when we apply the concept of perennial science. Aristotle was not pragmatic in developing his concepts, he was speculative as today science fiction authors are. That is why I refuse to call him a scientist, while *Heraclites*, his contemporary, was well an original scientist.

THE NEW NATURAL ORDER

I would like to elucidate some of the elements that both perennial philosophy and postmodern science share, as ingredients of a soup that today we came to call *holistic science*. As Fritjof Capra has shown in his bestseller *The Turning Point (1987)* and also in his books *The Web of Life (1997)* and *The Hidden Connections (2002)*, we are in the midst of a complete paradigm change in science which will *eventually declare wrong and obsolete* four hundred years of scientific error in the form of so-called 'exact,' Cartesian, reductionist science.

—See Fritjof Capra, The Turning Point (1987), The Web of Life (1997), The Hidden Connections (2002), and more recently, The Systems View of Life (2014).

My desire is to show that there are basically twelve, and probably more, ingredients and characteristics of holistic science that are presently more and more embraced, as we mature into new science which is of course just a modern vintage of perennial science. These twelve emanations or branches of the tree of knowledge remain still forbidden for most humans today because they follow the Devilish oversoul of the mass media, instead of following their own lucid inner wisdom.

NATURAL ORDER

Ancient Wisdom Traditions

Ancient traditional cultures and their scientific traditions, and what we today call perennial philosophy were holistic; they embraced *flow principles*, they looked at life as a *Gestalt*, and derived conclusions from the observation of the living and moving, not from the dead. Here are some of the most important of these traditions:

- Ancient Sumer
- Ancient Babylon
- Ancient Egypt
- Ancient Persia
- Ancient Greece
- Ancient Rome
- Ancient India
- Ancient China
- Ancient Japan
- Ancient Ottoman Empire (Ancient Turkey)

THE NEW NATURAL ORDER

Goethe's Color Theory

There was one genius in human science history, most of the time overlooked by our arrogant scientific pulpits, who was the real precursor of holistic science, at a time when everybody got Newtonian reductionism thrown over the head like a Cartesian mass-medicine. No, it was not Reich, while I always thought it was Reich, but just as a matter of timeline, there was *one* before him. It was the German lawyer, poet, philosopher and scientist Johann Wolfgang von Goethe (1749-1832).

Goethe, besides other scientific novelties, developed a color theory that was in flagrant contradiction with Newton's reductionist paradigm, and that is why Goethe was shunned by the mainstream science hierarchy not for decades, but for centuries. And Goethe knew why he had to oppose Newton! Though the merits of Goethe's color science, as advanced in his text *Zur Farbenlehre*, have often been acknowledged, it has been almost unanimously proclaimed invalid as physics.

How could Goethe have been so mistaken? Dennis L. Sepper shows that the condemnation of Goethe's attacks on Newton have been based on erroneous assumptions about the history of Newton's theory and the methods and goals of Goethe's color science. By illuminating the historical background and the experimental, methodological, and philosophical aspects of Goethe's work, Sepper shows that Goethe's color theory is in an important sense genuinely physical, and that, simultaneously as poet, scientist, historian, and philosopher, Goethe managed to anticipate important twentieth-century research not only in the history and philosophy of science, but even in color science itself.

—See Dennis L. Sepper, Goethe Contra Newton (1988), and Frederick Burwick, The Damnation of Newton (1986).

The Twelve Branches of the Tree of Knowledge

- Science and Divination

- Science and Energy

- Science and Flow

- Science and Gestalt

- Science and Intent

- Science and Intuition

- Science and Knowledge

- Science and Pattern

- Science and Perception

- Science and Philosophy

- Science and Truth

- Science and Vibration

Science and Divination

When I talk about *divination*, I include all possible devices, methods and traditions that are used for getting a glimpse of truth for decision-making, or potential outcome of specific events around a chosen developmental theme. Thus, divination can mean *astrology*, it can mean *Tarot* and it can mean *geomancy*, and it definitely also can mean using the *I Ching*.

Cartesian science never cared about explaining divination and why it works, while archetypal and transformational psychology, especially the Jungian branch of it, offered a pioneering and thought-provoking pathway for opening the depth of the psyche and its divinatory potential to the modern researcher or psychologist. One of the leading publications in this context is Sallie Nichols' *Jung and Tarot*.

—Sallie Nichols, Jung and Tarot (1986).

It is important in this context to realize that divination is not deterministic in the sense that 'the future is predetermined', while this assumption often appears to be repeated in vulgarized publications on esoteric sciences. The truth is that no diviner can ever predict 'the future,' as the future is simply an extrapolation of present thought content, and subconscious thought patterns, as well as emotional patterns.

What the diviner does is in fact scan the content of our unconscious and project this content into some or the other cognitive system that renders it visible and intellectually graspable. Hence, what divination explains is but the status quo of the asker, the person who comes to the diviner, with a particular question or project. While it is true there is a certain probability that the present state of consciousness perpetuates itself into the future, by extrapolation of its content on

THE NEW NATURAL ORDER

a timeline of events, this is no 'prediction' of the future, simply because the asker can change their content of consciousness *hic et nunc*.

This is why I developed, years ago, the idea of combining astrology and other forms of divination with what I came to call *Creative Prayer* as part of my *Life Authoring* self-coaching technique. The prayer technique is used as an add-on to the astrological consultation in the sense that it helps changing the present content of consciousness, after it has been rendered cognizable by the projective system of astrology. I learnt the technique basically from three books by Dr. Joseph Murphy, *The Power of Your Subconscious Mind*, *The Miracle of Mind Dynamics* and *Think Yourself Rich*.

> —Joseph Murphy, The Power of Your Subconscious Mind (1963), The Miracle of Mind Dynamics (1964) and Think Yourself Rich (2001).

The solution to the riddle of how divination works is contained in one single phrase of this book. Here it is:

> Remember that because your future is the result of your habitual thinking, it is already in your mind unless you change it through prayer.

NATURAL ORDER

— Dr. Joseph Murphy, The Power of Your Subconscious Mind, p. 165

What divination does, to repeat is, is to read our habitual and repetitive thought patterns, and extrapolate them on a virtual timeline into the future. This is, then, what is colloquially called 'predicting the future'. When you know what it's really about and how it is done, you see that it doesn't make sense, but can be understood as an oversimplification of the truth.

After all, if the future was predestined, as Calvinism assumed, Joseph Murphy and many other new thought authors would not have written their books; and they would not lecture as ministers and spiritual guides. They do it because they have realized that wrong beliefs about life and living are destructive and make for much of the misery we encounter in human lives, and in the world at large. Our mind is fragile in the sense that it can easily be manipulated by the mass media; worse, when fortune tellers, astrologers and diviners come along to pretend they are 'predicting the future,' the outcome may even be dangerous as their assumptions may by naive souls be taken as hypnotic spells that then may gain the power to realize as *self-fulfilling prophecies.*

THE NEW NATURAL ORDER

The reader may easily imagine where this can lead, and how much strife and turmoil this may produce in the lives of many humans around the world.

Murphy has seen it all around himself, and even in his own family, how people fall ill and even die without having to die, because of the suggestions they receive from others *in the form of hypnotic spells* wrapped in various forms, and also, unfortunately, in professional divination, when done by unspiritual, greedy and dishonest people.

And it's a fact, only to look at the Internet, what masses of scam artists are around in all those fields called esoteric, new age, mindpower and all the rest of it. When such accumulated power of irresponsible manipulative greed meets the *fragile and ignorant mind* of the 'man in the street,' then we can virtually predict disaster to happen.

This messy situation with people being misled by both their own beliefs and predictions received from others was one of the reasons that motivated Murphy and before him, Ernest Holmes, to write their books. The science that I have in mind when I put up the dichotomy science vs. divination is the so-called *Science of Mind*, also called *Religious Science*, as it was founded by Ernest Holmes in 1927, and expanded and commercialized in the 1960s by Dr. Joseph Murphy and Catherine Ponder, and others. I studied the *Science of Mind* thoroughly over the last twenty years; it clearly emphasizes the priority of mind over matter—spiritual monism—and also the priority of the present over the past and any form of predestination.

It is not known to many that the Bible is against both astrology and any form of fortune telling. For example, Deuteronomy 18: 9-12 affirms:

> Deuteronomy 18: 9-12
>
> 9 When thou art come into the land which the LORD thy God giveth thee, thou shalt not learn to do after the abominations of those nations.
> 10 There shall not be found among you any one that maketh his son or his daughter to pass through the fire, or that useth divination, or an observer of times, or an enchanter, or a witch.
> 11 Or a charmer, or a consulter with familiar spirits, or a wizard, or a necromancer.
> 12 For all that do these things are an abomination unto the LORD: and because of these abominations the LORD thy God doth drive them out from before thee.

When I came across these Bible quotes in 1991, I was first revolted! I found the Bible forwarded here a form of Christian fundamentalism that was against my convictions and spirituality. Yet I wanted to understand what the Bible meant here, what the deeper meaning was behind these admonitions. Thus I was asking 'how does the Bible relate to divining'? And why does it exhort us to be careful with it? To begin with, let me quote a mind-boggling example from Murphy's book *The Power of Your Subconscious Mind (1962/1982).*

> DR. JOSEPH MURPHY
>
> *How Suggestion Killed a Man*
> A distant relative of mine went to a celebrated crystal gazer in India and asked the woman to read his future. The seer told him that he had a bad heart. She predicted that he would die at the next new moon.
>
> My relative was aghast. He called up everyone in his family and told them about the prediction. He met with his lawyer to make sure his will was up-to-date. When I tried to talk him out of his conviction, he told me that the crystal gazer was known to have amazing occult powers. She could do great good or harm to those she dealt with. He was convinced of the truth of this.
>
> As the new moon approached, he became more and more withdrawn. A month before this man had been happy, healthy, vigorous, and robust. Now he was an invalid. On the / predicted date, he suffered a fatal heart attack. He died not knowing he was the cause of his own death.
>
> How many of us have heard similar stories and shivered a little at the thought that the world is full of mysterious uncontrollable forces? Yes, the world is full of

forces, but they are neither mysterious nor uncontrollable. My relative killed himself, by allowing a powerful suggestion to enter into his subconscious mind. He believed in the crystal gazer's powers, so he accepted her prediction completely.

Let us take another look at what happened, knowing what we do about the way the subconscious mind works. Whatever the conscious, reasoning mind of a person believes, the subconscious mind will accept and act upon. My relative was in a suggestible state when he went to see the fortune teller. She gave him a negative suggestion, and he accepted it. He became terrified. He constantly ruminated on his conviction that he was going to die at the next new moon. He told everyone about it, and he prepared for his end. It was his own fear and expectation of the end, accepted as true by his subconscious mind, that brought about his death.

The woman who predicted his death had no more power than the stones and sticks in the field. Her suggestion in itself had no power to create or bring about the end she suggested. If he had known the laws of his mind, he would have completely rejected the negative suggestion and refused to give her words any attention. He could have gone about the business of living with the secure knowledge that he was governed and controlled by his own thoughts and feelings. The prophecy of the seer would have been like a rubber ball thrown at an armored tank. He could have easily neutralized and dissipated her suggestion with no harm to himself. Instead, through lack of awareness and knowledge, he allowed it to kill him./

In themselves, the suggestions of others have no power over you. Whatever power they have, they gain because you give it to them through your own thoughts.

THE NEW NATURAL ORDER

You have to give your mental consent. You have to entertain and accept the thought. At that point it becomes your own thought, and your subconscious works to bring it into experience.

Remember, you have the capacity to choose. Choose life! Choose love! Choose health!

— Joseph Murphy, The Power of Your Subconscious Mind (1962/1982), pp. 29-30.

There is a difference between foolishly accepting any 'prediction' by an astrologer or fortune teller, or to use, for example, the I Ching, for decision making. The same is true regarding serious astrology; it is a question of professional ethics to *avoid being suggestive* in any way.

This is equally true for a serious Feng Shui consultant, Tarot expert, and even for paranormals who practice their profession within the rules of the unwritten ethical code set in Antiquity for all *Hermetic Sciences*. But from the side of the client, a certain level of emotional maturity is equally required! How many people die because they receive 'death sentences' from their physicians, taking for granted that the gods in white coats determine their destiny, and for the most part ignorant about the pitfalls, limitations and outright ignorance of Western medical science! There is a responsibility linked to every new piece of knowledge you learn and digest. This responsibility requires you to use the knowledge not with a foolish, immature or infantile mindset that takes everything for granted when it comes from a so-called 'authority'.

Science and Energy

When we look at the ingredients of holistic science, we have to begin with a notion that is primordial to all matter and visible phenomena. It was traditionally called cosmic life energy. What we have to ask, what kind of energy can that possibly be?

According to Albert Einstein's mass-energy equation ($e=mc^2$), kinetic energy in the universe results from an interdependence of mass and velocity. However, the energy meant in this equation is *not* the cosmic information field or zero-point field. This can be seen in the simple fact that kinetic energy is quantifiable, the primordial energy however not; as a result, the first can well be measured, the second, hitherto, not. For this more primordial 'energy,' other laws are applicable. For example, Einstein's speed of the light limitation while it is valid for matter, for large bodies, is not valid on the subatomic level, and it is not valid for the information field. Within the field or A-field, as Ervin Laszlo calls it, information flow it total and instantaneous, and this is so from one point of the universe to any other point, which may be light years away. The information will be transmitted instantaneously because of the *entanglement of particles*, a phenomenon explained either within the terminology of quantum physics or with the vocabulary created by Rupert Sheldrake, as *morphic resonance*.

> —See, for example, Lynne McTaggart, The Field (2002), Ervin Laszlo, Science and the Akashic Field (2004), William A. Tiller, Psychoenergetic Science (2007) and Conscious Acts of Creation (2001), Richard Gerber, A Practical Guide to Vibrational Medicine (2001), Rupert Sheldrake, A New Science of Life (2005). See also Valerie Hunt, Science of the Human Vibrations of Consciousness (2000) and Michael Talbot, The Holographic Universe (1992).

Though I well obtained interim answers to my initial quest to find out about the primordial energy, through Paracelsus, Swedenborg and Mesmer, through Reichenbach and Reich, through Burr and Lakhovsky, and also of course through Chinese Medicine, Feng Shui, Homeopathy, Bach Flowers and Hypnosis, through Huna and the wisdom of the

NATURAL ORDER

Essenes, I came to conclusive answers only after I got to understand quantum physics and modern research on the a-field, zero-point field or quantum vacuum, as well as morphic resonance. on the question, as conducted, for example, by Laszlo or Tiller.

At a time, some twenty years ago, when Western science was much more resistant to the idea of a *primordial energy field, or energy patterns*, Fritjof Capra courageously tackled the hairy problem in his book *The Turning Point (1982)*, discussing the controversial position of Wilhelm Reich, who termed the cosmic energy *orgone:*

> FRITJOF CAPRA
>
> It is evident that Reich's concept of bioenergy comes very close to the Chinese concept of ch'i. Like the Chinese, Reich emphasized the cyclical nature of the organism's flow processes and, like the Chinese, he also saw the energy flow in the body as the reflection of a process that goes on in the universe at large. To him bioenergy was a special manifestation of a form of cosmic energy that he called orgone energy. Reich saw this orgone energy as some kind of primordial substance, present everywhere in the atmosphere and extending through all space, like the ether of the nineteenthcentury physics. Inanimate as well as living matter, according to Reich, derives from orgone energy through a complicated process of differentiation. (Id., p. 378)

THE NEW NATURAL ORDER

When we try to find a *unified terminology* for the cosmic energy field, we need to make abstraction from the wrapper; language is a mere fold for content that is subject to observation.

Terence McKenna observes regarding the terminology used by tribal peoples to describe energetic phenomena that it metaphorically says basically the same as modern science. In *The Archaic Revival (1992)*, and with regard to the bioenergetic charge contained in plant substances used for religious purposes, McKenna writes:

> They [the natives] are the true phenomenologists of this world; they know plant chemistry, yet they call these energy fields spirits. (Id., p. 45)

But of course, McTaggart, Laszlo, Tiller or Gerber are not the last word. And let's not forget that despite these authors being recognized authorities in their field, this means only that we got some authoritative views, but not more. The last word of mainstream Western science regarding the integration of the cosmic energy field is not out until this day. And that means that—so far—this science operates around the main parameter of the universe. Which is quite of an elegant workaround after all. I haven't heard of a pianist who can play without a piano. Dora van Gelder expresses this beautifully when she says, in her book *The Real World of Fairies (1977/1999)*:

> We live in a world of form without understanding the life force beneath the forms.

Science and Flow

When we look at the ingredients of holistic science, we have to begin with *flow*. There is an important difference between a static and a dynamic science concept.

The Cartesian science concept was static. Instead of looking at the living and moving substance, it vivisected dead corpses to gain insights about life.

Why was that so? Cartesianism disregarded flow, while ancient traditional cultures and their scientific traditions, and what we today call perennial science, were holistic; they embraced flow principles. They looked at life as a *Gestalt*, and derived conclusions from the observation of the living and moving, not from the dead.

Our own holistic science tradition probably started not with Aristotle, but with Heraclites.

THE NEW NATURAL ORDER

> —See, for example, Richard Geldard, Remembering Heraclitus (2000) and Charles H. Kahn (Ed.), The Art and Thought of Heraclitus (2008).

Heraclites' most important principle in nature was flow, the *flow principle* as we would say today an as it was rediscovered in systems theory, the science of living systems.

The most pertinent general information on the flow principle I found in the books of Fritjof Capra, while I know that there is much more specialized literature about this subject. Yet Capra's books are really instructive in that they contain all the references needed to research these publications. In most cases, it suffices to read Capra's very well-written summaries to get the picture. His elucidations provide a bird perspective on systems theory and the intricacies of living systems, and from there you can go deeper, using the references as guides.

> — Fritjof Capra, The Tao of Physics (1975/2000), The Turning Point (1987), The Web of Life (1997), The Hidden Connections (2002) and Steering Business Toward Sustainability (1995). See also Peter Fritz Walter, Fritjof Capra and the Systems View of Life: Short Bio, Comments & Book Reviews, Great Minds Series, Vol. 4 (2015/2017).

NATURAL ORDER

SCIENCE AND GESTALT

As I have shown in my reviews of some of the lesser known books by Wilhelm Reich, this scientist's conceptual framework had firmly embodied the *Gestalt*. Reich's genius as a scientist was his gift of observation, and his particular talent to see not single elements of a process, but the *whole* of the process. Reich was in this respect really different from the main bunch of his Cartesian-minded professional colleagues. In our days, Reich would probably be considered as one of the leading-edge scientists.

Generally speaking, when we observe living processes, we can either put our focus on single elements, or the substance, or we can focus on the *process*, and the *form*. Both form and substance are present in living systems.

Our culture has created the line as a symbol for evolution. However, the straight line is an *artificial construct*, inexistent in nature, a purely mental achievement, while evolution is cyclic. It allows the line only in combination with the circle, so as to say, resulting in the *spiral*. Merriam-Webster's Dictionary defines the spiral as *relating to the advancement to higher levels through a series of cyclical movements.*

THE NEW NATURAL ORDER

The curving movement of the spiral is what it has in common with the circle; the increase or decrease in size of the spiral is a function of its moving upward or downward.

Interestingly enough, the spiral is by far the dominating form to be found in nature, and in all natural processes. It is a symbol or Gestalt for evolution in general. Life is coded in the spiraled double-helix of the DNA molecule. The spiral is the expression of the periodic, systemic and cyclic development that is in accordance with the laws of life. The progression of the spiral shows that it always carries its root, however transporting it through every cycle onto a higher level or dimension; whereas the line leaves its root forever. All towers of Babel are manifestations of the line; they are linear and are created by linear thought structures.

True growth is always cyclic and spiraled, and nonlinear.

On the subject of bringing in Gestalt thinking in the logic of healing, Manly P. Hall, in his book *The Secret Teachings of All Ages (1928/2003)* writes about Paracelsus and states:

Paracelsus discovered that in many cases plants revealed by their shape the particular organs of the human body which they served most effectively. The medical system of Paracelsus was based on the theory that by removing the diseased etheric mumia from the organism of the patient and causing it to be accepted into the nature of some distant and disinterested thing of comparatively little value, it was possible to divert from the patient the flow of the archaeus which had been continually revitalizing and nourishing the malady. Its vehicle of expression being transplanted, the archaeus necessarily accompanied its mumia, and the patient recovered. (Id., p. 347)

It was Gestalt considerations and the insight that nature is basically an *assemblage of patterns*, and not of randomly arranged matter that recently led a number of researchers to corroborate the age-old assumption that our universe is *holographic*, and thus programmed in holographic patterns that are all mutually interconnected. Ervin Laszlo writes in his remarkable study *Science and the Akashic Field (2004)*:

> In a holographic recording—created by the interference pattern of two light beams—there is no one-to-one correspondence between points on the surface of the object that is recorded and points in the recording itself. Holograms carry information in a distributed form, so all the information that makes up a hologram is present in every part of it. The points that make up the recording

of the object's surface are present throughout the interference patterns recorded on the photographic plate: in a way, the image of the object is enfolded throughout the plate. As a result, when any small piece of the plate is illuminated, the full image of the object appears, though it may be fuzzier than the image resulting from illuminating the entire plate. (Id., p. 55)

Science and Intent

It was only fairly recently that modern science began to ask if, and in how far, intent, human intention, impacts upon matter, or even may contribute to changing matter?

This very question would have been judged insane still a few decades ago by the bulk of modern scientists. The question of how intention impacts upon matter is generally asked in the context of what is called 'Mind and Matter Interaction,' and this is itself a topic related to consciousness research.

I have for the first time heard of this new pathway in modern research through the film *What The Bleep Do We Know!?* and its later *Rabbit Hole Edition* and was then reviewing, just a few months ago, the mind-boggling presentation by William A. Tiller,

NATURAL ORDER

Stanford University Emeritus entitled *Conscious Acts of Creation (DVD, 2004)*.

Shortly thereafter I found in Chapter 8 of the book *The Conscious Universe (1997)*, by Dean Radin, entitled *Mind-Matter Interaction* the following interesting remark; it's not the answer yet, but a well formulated question:

> Does mental intention affect the physical world? In a trivial sense, the answer is obviously yes. An automotive engineer imagines a new way to build a car, and several months or years later it appears. This transformation from mental into physical is not considered remarkable because the sequence of events is well understood.
> But a similar question can be asked that is no longer self-evident: does mental intention directly affect the physical world, without an intermediary? This question concerning the ultimate role of the human mind in the physical world has intrigued philosophers for millennia. Indeed, the concept that mind is primary over matter is deeply rooted in Eastern philosophies and ancient beliefs about magic. For the past few hundred years, such beliefs have been firmly rejected by Western science as mere superstition. And yet, the fundamental issues remain as mysterious today as they did five thousand years ago. What is mind, and what is its relationship to matter? Is the mind caused, or is it causal?

Answers are now given by a number of scientists, among them Dr. Tiller. Based upon years of detailed research, Dr. Tiller has amassed convincing experimental data showing that in seemingly the same cognitive space, basic chemical reactions and material properties can be strongly altered by human intentions. He says, we are all capable of performing what we think of as miracles. In *Conscious Acts of Creation*, Dr. Tiller explains these findings in clear, understandable language and supports it with just enough math and physics to deeply move just about everyone.

The exciting bottom line is that it appears now real that these findings and the new technologies Tiller has developed, are capable of catalyzing human intention in the process of changing matter!

While Dr. Tiller's research investigates the impact of intent on mechanics, simple machines and their workings when under the influence of conscious intent, I am actually interested to use this methodology to explain shamanism, and particularly shamanic healing.

It is on the same line of reasoning when I say that shamans use intent for altering consciousness. For example, it can be asked how intent comes into play during shamanic healing, and how, in detail, a shaman's conscious intent impacts upon the consciousness matrix of the receiver?

I namely argue that the shaman does not impact directly on the consciousness of his client but by using *entheogenic plants as a receptor, amplifier and emitter platform* for thought content, and for intention.

Science and Intuition

As early as in high-school I was awake, critical and suspicious why my science teachers were so painstakingly discarding intuition out from the core scientific method and paradigm. I simply thought they were feeble-minded, which is probably why I never learnt much science. But I don't regret it, because what I would have learnt, as I know today, would have been wrong—completely wrong.

In Antiquity, intuition truly had its place, and this is now beginning to dawn on the cutting edge of modern science, *as intuitive diagnosis of illness*, an idea that would have sounded science-fiction just a few decades ago, begins to be seriously recognized within modern medicine.

THE NEW NATURAL ORDER

Caroline Myss, in her practice, has shown to intuitively diagnose illness with a more than eighty percent accuracy, which is simply phenomenal. In her contribution to Russell DiCarlo's science reader, *Towards A New Worldview (1996)*, she says she relates intuition to the vital energy.

In her view, intuition is one of the many manifestations of the *ch'i* energy or cosmic information field. She writes:

> The human energy field shouldn't be called that at all, but since we call it that, let's define it very clearly. It's better understood as an information center because that's what it is. And that's where you store all your messages. That's where you store all your faxes. That's where you warehouse everything. Your responses to everything and everyone, all your fear - everything - is stored in your energy field. Your responses form patterns that influence your electromagnetic circuitry. This dictates a quality control signal that influences the creation and quality of cell tissue'. (…) Energy is intelligent. It is alive. It is information - energy is information. It is one and the same thing.
>
> —Carolyne Myss, in: Russell DiCarlo (Ed.), Towards A New Worldview (1996), pp. 136-145.

Barbara Brennan, in her contribution to the same science reader, emphasized that all our thoughts and emotions impact upon the energy field in which we are woven. As such, what we *perceive and intuit* is transmitted to us by energy, because energy is information.

> In between the structured layers of the field is a bioplasma-like energy that simply flows along the lines of the structured field pattern. It's the energy that flows along the lines of the structured field pattern that changes very fast with thoughts and emotions, not the structured pattern itself. For example, if you stop yourself from feeling something, it will stop the flow of energy in the field. And if you experience the feeling, the energy will be released. There is a direct correlation. There are even correlations between the energy field and the part of the brain you are thinking with. As you change your thought patterns, the patterning of the field changes. (Id., p. 147)

The idea of intuition being a variant of *direct perception* is old, very old. It was an idea very dear to Pythagoras. Manly P. Hall writes in *The Secret Teachings of All Ages (1928/2003)*:

> Pythagoras defined knowledge as the fruitage of mental accumulation. He believed that it would be obtained in many ways, but principally through observation. Wisdom was the understanding of the source or cause of all things, and this could be secured only by raising the intellect to a point where it intuitively cognized the invisible manifesting outwardly through the visible, and thus became capable of bringing itself en rapport with the spirit of things rather than with their forms. (Id., p. 197)

THE NEW NATURAL ORDER

Manly P. Hall reports that John Stuart Mill believed in intuition and reason to be the *two superior modes of apprehending reality*, and that they are higher states of the mind compared to mere sensory perception.

> John Stuart Mill believed that if it is possible through sensation to secure knowledge of the properties of things, it is also possible through a higher state of the mind—that is, intuition or reason—to gain a knowledge of the true substance of things. (Id., p. 34)

Vidette Todaro-Franceschi, in her book *The Enigma of Energy (1991)*, has asked the question 'What, Exactly, is Nature?' Referring to historian and philosopher of science R. G. Collingwood, she writes that there are *three periods* in the development of the idea of nature, which she sees coincidentally reflect the ideas of energy. Then, on the subject of the third period in the developmental view of nature, she notes:

> I believe that during this period the idea that energy was an autonomous existent contributed to the shift in focus. It became vaguely evident that change was inherent in various things; that is, it was recognized that change could occur without the provocation of external forces or efficient causes. Collingwood identifies the idea of a 'rhythmical pattern' with the modern view of nature and acknowledges that the new physics theories are partly

responsible for this notion. But the rhythmical patterns we now know to exist in nature also seem to denote an inner principle of change, or an Aristotelian 'that for the sake of which,' originally expressed by the ancient Greeks. So one might say we have come full circle. In conjunction with this new take on an old idea that was present in both Eastern and Western antiquity is the increasing awareness that intuition plays a significant role in scientific discoveries. As the historical background of the idea of energy attests, intuitive ways of knowing have been crucial to the development of scientific ideas throughout history. Many individuals knew things, such as the energy conservation doctrine, without being able to empirically verify them. In other words, intuitive ways of knowing seem to have led / us in the right direction long before we were capable of scientifically validating what we somehow knew to be so. Subjectivity and subjective ways of knowing, such as intuition, have become as vital to our understanding as objectivity and empirical ways of knowing. In this modern view of nature humankind has once again come to be recognized as being part of nature, rather than outside of it. (Id., pp. 125-126)

Science and Knowledge

Is science per se a form of knowledge-gathering? Or is it rather the opposite, a system that inhibits us from gaining true knowledge? These questions guide us toward the insight that all depends on our definition of science.

THE NEW NATURAL ORDER

Scientia, the Latin word, originally meant indeed knowledge. But has our traditional science led us toward *true knowledge*, true knowledge of the nature of life, the wisdom of living, the art of togetherness and of peace?

I believe it has rather done the contrary, at least from the point it was transformed in a Cartesian reductionism, a split self of its original integrated whole, that it still was during Antiquity. In our Bible, which is for most people in Western traditions a true guide book, knowledge was in the Genesis considered as dangerous and it was explicitly forbidden to collect the fruits of the 'tree of knowledge.' The serpent, which in ancient traditions was always considered a consort of knowledge, was acting counter to this nonsensical prohibition and liberated man from the constriction imposed by a jealous, power-hungry persecutor-god.

And indeed most people, quite instinctively, because of their Christian upbringing, take the position of the serpent-killer, not the serpent-friend, and thus automatically become knowledge-hostile.

I contend that the larger part of Western culture and society is deeply knowledge-hostile. Even in today's pretendedly so enlightened international intelligentsia, knowledge is only accepted once it is promoted by a well-known academic authority and accepted on a larger scale—this acceptance being valid only when the majority among the knowledge-bearers are academia, and thus accredited at leading universities.

All other kind of knowledge is ruthlessly discarded out, and often those who bear the knowledge as well. Which is not democratic, nor is it enlightened.

The puzzle of true knowledge is a web-like structure, not a hierarchy of principles. It's nonlinear, not linear. It consists of many elements that are structurally related. When science cares about true knowledge, it first of all cares about *nature*, the true and only knowledge-giver in the universe. I will quote in this paragraph a few examples of knowledge that is not considered valid knowledge under our present science paradigm but that I consider as true knowledge, while I am not a clairvoyant myself.

But when science is primarily observation, then clairvoyant observation must be included in our plethora of scientific observations of nature.

THE NEW NATURAL ORDER

When I include this knowledge here, I do it in fair appreciation of clairvoyants such as Dora van Gelder or Charles W. Leadbeater, whose life stories I have studied, and in consideration of their high reputation as having been honest and scientific-minded people who received high regard and even admiration from the side of both academia and their contemporaries.

Now, first of all, what is clairvoyance? Dora van Gelder writes in *The Real World of Fairies (1999)*:

> The fact is that there is a real physical basis for clairvoyance, and the faculty is not especially mysterious. The power centers in that tiny organ in the brain called the pituitary gland. The kind of vibrations involved are so subtle that no physical opening in the skin is needed to convey them to the pituitary body, but there is a special spot of sensitiveness just between the eyes above the root of the nose which acts as the external opening for the gland within. (Id., p. 4)

> So the pituitary gland is certainly very much alive and important in human beings. And it certainly has this use for receiving very fine vibrations from a world of things which are subtler than anything we know. (Id., p. 5)

And what does modern science know about the *spirits of nature, and the fairy worlds?* Nothing. And because it knows nothing, it says that 'these things' do not exist. Very intelligent indeed. I think that modern science is a special form of religious fundamentalism that has not yet been identified as such. The spirits of nature, shunned so much by Christian fundamentalism, that was the predecessor of modern science, and reborn now in the course of the New Age movement, and the revival of the folk lore of fairies, as it was, for example, rediscovered by Dr. Evans-Wentz in his remarkable study *The Fairy Faith in Celtic Countries (1911/2002)*, and observed by clairvoyant Dora van Gelder in her book *The Real World of Fairies (1977/1999)*, have certain well-defined characteristics and they are quite distinct of human beings. Leadbeater explains:

Charles W. Leadbeater

> We might almost look upon the nature-spirits as a kind of astral humanity, but for the fact that none of them - not even the highest—possess a permanent reincarnating individuality. Apparently therefore, one point in which their line of evolution differs from ours is that a much greater proportion of intelligence is developed before permanent individualization takes places; but of the stages through which they have passed, and those through which they have yet to pass, we can know little. The life-periods of the different subdivisions vary greatly, some being quite short, others much longer than our human lifetime. We stand so entirely outside such a life as theirs that it is impossible for us to understand much about its conditions; but it appears on / the whole to be a simply, joyous,

irresponsible kind of existence, much such as a party of happy children might lead among exceptionally favourable physical surroundings. Though tricky and mischievous, they are rarely malicious unless provoked by some unwarrantable intrusion or annoyance; but as a body they also partake to some extent of the universal feeling of distrust for man, and they generally seem inclined to resent somewhat the first appearances of a neophyte on the astral plane, so that he usually makes their freaks, they soon accept him as a necessary evil and take no further notice of him, while some among them may even after a time become friendly and manifest pleasure on meeting him.

—Charles W. Leadbeater, Astral Plane (1997), p. 61.

The Adept knows how to make use of the services of the nature-spirits when he requires them, but the ordinary magician can obtain their assistance only by processes either of invocation or evocation—that is, either by attracting their attention as a suppliant and making some kind of bargain with them, or by endeavouring to set in motion influences which would compel their obedience. Both methods are extremely undesirable, and the latter is also excessively dangerous, as the operator would arouse a determined hostility / which might prove fatal to him. Needless to say, no one studying occultism under a qualified Master would ever be permitted to attempt anything of the kind at all. (Id., pp. 61-62)

And regarding medical science, the picture is not much different. It ignores more than it knows, it shuns and discards more than it embraces and recognizes in its medical paradigm. Dr. Alberto Villoldo, in his book *Shaman, Healer, Sage (2000)*, remarks:

> Until fifty years ago, going to a doctor was more dangerous to your health than staying home and letting your body-mind take its own course. (Id., p. 13)

I think Villoldo is rather optimistic here; in my view this is still true today. You may have a little carcinoma, a little tumor, or even a big tumor, but anyway, that's not a reason to get a death sentence pronounced by your medical executioner and a Nazi torture called chemotherapy to get you back to system-conformity and brave emotional dullness! Your body revolts with a tumor because you have abused it with conforming, probably over some decades, with a dead culture that is essentially pro-bomb and anti-body. Because you have denied your particular perversity and thought it was not politically correct. That's why people get cancers, not because of what medical business writes in their propaganda leaflets.

You can have all true information about cancer from alternative cancer specialists such as *Simonton & Simonton*, who are medical doctors of a different kind because they did not sell their souls to medical business.

THE NEW NATURAL ORDER

—See O. Carl Simonton, Getting Well Again (1968/1992).

Villoldo recognizes the power of the mind over the body when he recalls his own childhood:

ALBERTO VILLOLDO

The very real effects of the mind on the body have been confirmed by research. In a sense, we all became experts at developing psychosomatic disease very early in life. At the age of six I could create the symptoms of a cold in minutes if I did not want to go to school. Psychosomatic disease goes against every survival instinct programmed into the body by three hundred million years of evolution. How powerful the mind must be to override all of these survival and self-preservation mechanisms. Imagine if we could marshal these resources to create psychosomatic health!

—Alberto Villoldo, Shaman, Healer, Sage (2000), p. 13.

It seems that there is in our culture always a form of 'official knowledge' that is put forward to kill off all *real* knowledge. In present times, this killer app is *biotechnology*, commonly called *gene technology*. In similar ways as Fritjof Capra, who is one of the most explicit opponents to genetic determinism, Alberto Villoldo, in his book *Mending the Past and Healing the Future with Soul Retrieval (2005)* has lucidly analyzed the myth of genetic imperialism pervading our present-day culture.

NATURAL ORDER

He comes to the following conclusion, which is an empathetic statement for deep ecology. I agree as there cannot be a culture of knowledge that destroys its own roots.

ALBERTO VILLOLDO

Our behavior is a form of matricide, in which the child of nature - the human - is killing its own mother. To protect herself, nature is beginning to reject us: Water supplies are drying up, new plagues are infecting the planet, and the earth is beginning to respond to us as an undesirable life form. We're becoming a flea on the tail of a dog, a germ that will be annihilated by the immune system of the planet.

All this comes at a time when medicine feels newly empowered by our discoveries of the secret of life. When Watson and Crick discovered the DNA code, we suddenly converted to a new scientific faith, and antimicrobial medicine became supplemented by genetics. We now believe that risk factors inherited from our parents and ancestors through our genes predispose us to how long we're going to live (and how well), what illnesses we're going to get, how we're going to heal, and how we're going to age. We've devised tests to tell us from birth what genetic risks we've inherited, and we race to find cures from the same DNA strands that we use to predict our future. Genetic markers, nanotechnology, and other tools of the biotechnology industry promise us healthier and longer lives.

But this is just a new trick for an old dog, because biotechnology is still looking for ways to fix, correct, and kill at an even subtler molecular level. We've simply added more precision and skill to the attack, while what we should be doing is seeking harmony with nature, both inside and out.

—Alberto Villoldo, Mending the Past and Healing the Future with Soul Retrieval (2005).

A true culture of knowledge will put a preference and urgency agenda over all other seeming priorities for protecting the very base layer of life – our *Mother Earth*.

Science and Pattern

I have stressed in all my publications the importance of understanding the nature of our universe as a basically *patterned universe*; on the basis of this insight, I am addressing scientists to focus on patterned intelligence, or patterned organization when we observe nature.

What are patterns? I began identifying the perennial pro-life patterns in living by firstly invalidating the fake principles that mainstream Western science declares to be the founding concepts of our universe. To put it more precisely, there was actually nothing to invalidate: I found that these alleged principles were but *intellectual assumptions* and thus *simply invalid* as founding principles of life. At the same time, diligent study of the I Ching and almost daily use of it for divination during more than twenty years distilled in me an intuitive understanding of the *real and valid patterns that are inherent in all living*. I therefore simply called them *patterns of living*.

Let me first of all explain why I use the term *patterns*, consciously deciding to discontinue the use of the term *principles*. I indeed think that here we are facing a key point that marks the essential difference between death science and life science.

A *pattern* is a set of things, a certain arrangement I can make out in the complex scheme of reality. It is something I can *observe*. A pattern can be fix or it can be changeable. It can be static or dynamic.

By contrast, a *principle* typically is the beginning of a down-hierarchy. It's a top-something in a kind of up-to-down order. It is *not* something I can observe. Its reality is merely intellectual, the outcome of a conclusion I draw in my rational mind *after* observing nature. A principle thus contains my observer point or my judgment about reality.

THE NEW NATURAL ORDER

Death science looks at life through the glasses of principles it has set before it was going to observe. It is essential blind, and it proceeds by imposing characteristics upon nature.

Western science is death science. Traditionally, it gained its first conclusions about life by vivisecting cadavers, not by observing the moving changes of living. It is, and remained, a cadaver science that is far removed from the changing patterns of reality.

Life science looks at life without any set principles or assumptions and observes the dynamic patterns or changes in the texture of life. It is a science that since its start in China, around five thousand years ago, was interested in *life*, and thus drew conclusions from life, and not from death.

Traditional Chinese science is *life science*, one branch of this very large body of science and philosophy being *Feng Shui*.

The I Ching is based upon *life science*, and is perhaps the highest condensation of it. Needless to add that, as such, it is non-judgmental and thus bears *no moralistic judgments about human behavior.* It looks at human behavior in exactly the same way it looks at all life patterns, and sees the changing nature of it before all. I am angry at twelve twenty and hungry at twelve thirty.

In his book *The Web of Life (1997)*, Fritjof Capra explains the importance of pattern when he explores into the meaning of self-organization, which is one major characteristic pattern of living systems:

> To understand the phenomenon of self-organization, we first need to understand the importance of pattern. The idea of a pattern of organization—a configuration of relationships characteristic of a particular system—became the explicit focus of systems thinking in cybernetics and has been a crucial concept ever since. From the systems point of view, the understanding of life begins with the understanding of pattern. (Id., p. 80)

When inquiring what patterns are, we need to change our basic setup of scientific investigation. Capra explains:

> In the study of structure we measure and weigh things. Patterns, however, cannot be measured or weighed; they must be mapped. To understand a pattern we must map a configuration of relationships. In other words, structure involves quantities, while pattern involves qualities. (Id., p. 81)

This really involves a *radical change in scientific thinking* because traditionally Cartesian science was quantity-based and measure-oriented, while systemic science is quality-based and relationship-oriented, a truth that Capra exemplifies when looking at the properties involved in the scientific focus of both static and systemic science theory:

> Systemic properties are properties of pattern. What is destroyed when a living organism is dissected is its pattern. The components are still there, but the configuration of relationships among them—the pattern—is destroyed, and thus the organism dies. (Id.)

The next important point to understand how nature 'thinks' is the cell's metabolism, the network that serves recycling. Capra succinctly elaborates in his book *The Hidden Connections (2002)*:

> When we take a closer look at the processes of metabolism, we notice that they form a chemical network. This is another fundamental feature of life. As ecosystems are understood in terms of food webs (networks of organisms), so organisms are viewed as networks of cells, organs and organ systems, and cells as networks of molecules. One of the key insights of the systems approach has been the realization that the network is a pattern that is common to all life. Wherever we see life, we see networks. (…) The metabolic network of a cell involves very special dynamics that differ strikingly from the cell's nonliving environment. Taking in nutrients from the outside world, the cell sustains itself by means of a network of chemical reactions that take place inside the boundary and produce all of the cell's

components, including those of the boundary itself. (Id., p. 9)

But the most revolutionary finding is that our usual habit of dissecting parts of a whole for further scrutiny and scientific investigation does not work with living systems. Why is this so? Capra pursues in *The Web of Life (1997)*:

> Ultimately—as quantum physics showed so dramatically—there are no parts at all. What we call a part if merely a pattern in an inseparable web of relationships. Therefore the shift from the parts to the whole can also be seen as a shift from objects to relationships. (Id., p. 37)

My hypothesis is that Western culture has never until now applied what I came to call the *Eight Dynamic Patterns of Living* and that it therefore is at the border of chaos, destruction or another kind of worldwide catastrophe, suffering from a schizoid mindset, the perversion of *love* into hate and sadism, rampant violence, the impudent slaughtering of minorities, famines that could easily be avoided, and generally a total lack of genuine spirituality which, by itself, already makes for a large part of the depression and psychosomatic disorders many consumers in postmodern international culture are suffering from. What I say is that the *Eight Dynamic Patterns of Living* have been respected and applied by all major tribal cultures including the North American Indians, and that therefore they have lived, and live, peacefully. With peacefully I do not mean an artificial Western peace concept which is complete nonsense as it is stuck and rigid, but a *dynamic peace continuum* that allows little fights and small wars to happen, as required by the dynamics of *yin* and *yang*, but that is so balanced that it will never trigger a major and global destruction.

The fact that our global industrial culture is at the trigger of this destruction in all possible ways, economical, social, health-wise, military-wise, ecological, and other ways, shows that the continuum balance that the *eight patterns* give is completely lacking in modern society's philosophy, science, military policy, diplomacy, politics and strategy. We all have consciously or unconsciously contributed to bring about the *emotional plague*, symbolized by the atomic bomb's mushroom.

Thus, the *eight patterns* could be taken as a guide concept and implemented in a new kind of lifestyle to be worked out as part of our presently evolving postindustrial global culture. That is the basic idea. I think that the *eight patterns* are tremendously useful as a base layer for establishing the ground principles of a new peaceful society, instead of beginning with Adam and Eve and go time and again though all anthropological material. I have actually done this and there is no more novelty in this. The *eight patterns* cover all spheres of life and living.

Science and Perception

Cartesian Science was *reductionist* in that it was limited to *sensory perception*, shutting out everything from scientific observation that could not be grasped with the five senses. By definition, thus, extrasensorial and multi-sensorial perception was simply discarded out from scientific scrutiny and relegated to 'mysticism,' 'imagination,' 'daydreaming' or somnambulism, to a point to actually label people insane who do have a complete range of natural perception. This arrogant science paradigm has not only created havoc in its own culture, but it definitely has been paradigmatically at the origin of large-scale genocide of native cultures.

That is why I call Cartesian science simply a cadaver science and a murder science. It can kill effectively, but it cannot heal. It can only destroy, but not create, dissect but not integrate, separate but not unite. It's simply perverse as all those who work for it with their eternal male hubris and their backpack of Oedipal hangups.

It has been the foundation of massacres against tribal populations for centuries and centuries and it has absolved the intentional murder of uncountable animals that it cruelly tortured and dissected in their lifeless and aseptic laboratories.

Alberto Villoldo, in his book *Shaman, Healer, Sage (2000)* distinguishes between *rule-driven*, *concept-driven* and *perception-driven* societies. While this distinction may not be clear-cut and cover all cases, it is a good and practically useful guideline that shows what could be called the *predominant orientation* of a given culture and its intelligentsia. Alberto Villoldo writes:

> We are a rule-driven society that relies on documents such as the Constitution, the Ten Commandments, or laws passed by elected officials to bring order to our lives. We change precepts (rules or laws) when we want to change the world. The ancient Greeks, on the other hand, were people of the concept. They were interested not in rules but rather in / ideas. They believed that a single idea could change the world and that there was

nothing as powerful as an idea whose time had come. Shamans are people of the percept. When they want to change the world, they engage in perceptual shifts that change their relationship to life. They envision the possible, and the outer world changes. This is why a group of Inka elders will sit in meditation envisioning the kind of world they want their grandchildren to inherit. (Id., pp. 9-10)

The predominant orientation of native peoples around the world is perception, not just perception but *direct perception*. Western science has very little understood so far what immediate, direct or primary perception is. Direct perception circumvents the judgment interface of the neocortex and thus short-circuits the rational mind. It is connected with the reptilian brain and the limbic system.

Dr. Villoldo calls it *primary perception*, for good reason as it was surely the primal form of perception in the run of human evolution. Its use decreased through settlement, domestication and human civilization, while in the shamanic world this primary form of perception is still the rule—at least for the shamans themselves.

Now, what is direct perception? It is as difficult to define perception as it is to define life. You can call it total unity with all-that-is, fusion with the object, an absence of the *second observer.*

> —Commonly, the expression 'secondary observer' connotes the judgment interface, our morality, the conditioning we have received and that impacts upon and distorts our perception of reality. The primary observer is another expression for what Transactional Analysis (TA) is called the inner adult. Accordingly, the secondary observer is called by TA the inner critic. It's not a very useful instance and is created artificially through a hypertrophy of the intellect and an overbearing superego. In simple words, it's accumulated shame or a shame-based identity.

A person who perceives reality totally could be called, for example, a totally conscious direct observer. Dr. Villoldo writes in *Shaman, Healer, Sage (2000):*

> To practice primary perception shamans have developed a kind of 'common sense' that bridges all of the senses. They are able to taste fire, to touch the fragrance of a flower, and to smell an image. They attain immediate perception before an experience is divided among the senses, an ability known as synesthesia. This blending of sensory modalities seems strange only to those who have distanced themselves from a direct, primordial experience of the natural world. (Id., p. 116)

I think it is possible to train our direct perception capacities in developing consciousness that apprehends reality beyond our five senses; this means to train ESP, extrasensorial perception and MSP, multisensorial perception. Extrasensorial perception is in our society discussed among the greater topic of parapsychology or psychic research, while multisensorial perception, as Dr. Villoldo remarks, is commonly associated with shamanism and synesthesia.

Synesthesia has been reported and documented, to my knowledge, from two famous musicians, Olivier Messiaen and Alexander Scriabin. Messiaen was commonly talking about colors in his music, specific colors that he could identify when the music was performed. He also claimed that his music was more colorful than, for example, the music of Johann Sebastian Bach. Scriabin's sense of synesthesia was vastly commented upon by his biographers.

Dr. Villoldo quotes an interesting passage from *Phenomenology of Perception (1945/1999)*, by Maurice Merleau-Ponty that claims synesthesia to be our primary mode of sensorial input.

> As the philosopher Maurice Merleau-Ponty wrote in Phenomenology of Perception, 'Synesthetic perception is the rule, and we are unaware of it only because scientific knowledge shifts the center of gravity of experience, so that we have unlearned how to see, hear,

THE NEW NATURAL ORDER

and generally speaking, feel, in order to deduce, from our bodily organization, the world as a physicist conceives it, that we are to see, hear and feel.' (Id.)

This is a classical example for the fact that our science is a *death science* that murders, shuts out, discards and ignores more in the universe than it embraces and knows about. And not only that. It also trains and conditions young citizens to perceive the world in a limited way, when it infiltrates into the school system. And in so far it mutilates the perception of our children. I do really not know what our science is good for other than for building fridges, airplanes, televisions and light bulbs? It has been good for providing us comfort and safety, but it has deprived us from, and shielded us against, most of the living world.

We can only hope that our fake science learns from shamanism and generally from the wisdom of ancient cultures and native peoples – who know what is *true science*.

Science and Philosophy

There is an important terminological clarification to be made regarding the terms *science*, on one hand, and *philosophy*, on the other. Traditionally, in Western culture philosophy was connoted with more or less vague assumptions about life, or a certain life program, and associated with, or even used as a synonym of, *Weltanschauung*. Science, by contrast, in the Western science tradition, was understood, as I pointed it out already, as Cartesian reductionism—a regard on the world that pretended to be exact, objective and methodologically sound, while it was clearly shutting out more from this world than it admitted in its residue paradigm of science.

I am using these terms in the precisely opposite sense, in the sense namely that is in accordance with the oldest of science traditions, the hermetic tradition, and perennial philosophy. In this sense, what was called *philos sophia* (Love for Knowledge) in Antiquity was the header notion for science, whereas philosophy in the sense it was used during the last four hundred years simply would have to be called *speculation*. In other words, applying the old holistic science concept of Antiquity, our modern science would represent but a tiny portion of that cake …

Manly P. Hall, in his book *The Secret Teachings of All Ages (1928/2003)*, observes:

THE NEW NATURAL ORDER

> Among the ancients, philosophy, science, and religion were never considered as separate units: each was regarded as an integral part of the whole. Philosophy was scientific and religious; science was philosophic and religious; religion was philosophic and scientific. Perfect wisdom was considered unattainable save as the result of harmonizing all three of these expressions of mental and moral activity. (Id., p. 253)

So-called 'exact science' in the old Cartesian meaning of the expression is not science, but reductionism. So the wisdom here is to remain open and flexibly intelligent so that the observer can shift and move and change—and only then a truly scientific approach is granted. As Fritjof Capra writes in his book *The Web of Life (1997)*:

> What makes it possible to turn the systems approach into a science is the discovery that there is approximate knowledge. This insight is crucial to all of modern science. The old paradigm is based on the Cartesian belief in the certainty of scientific knowledge. In the new paradigm it is recognized that all scientific concepts and theories are limited and approximate. Science can never provide any complete and definite understanding.

While my view was still a few years ago considered a minority opinion, it is now more and more recognized as the correct view, and it is currently developing into the new mainstream view of new science. What is this new science? It is mainly a renaissance, not an original new creation of modern minds, but this rebirth of perennial philosophy in a new garment is enriched by the irrevocable discoveries in quantum physics, and thus got a foundation that cannot be discussed or rationalized away anymore. And eventually, this science begins to recognize and acknowledge the fact that life is coded in holistic patterns that constantly change and evolve, and not as a hierarchic pyramid of stiff and eternal principles. Capra writes in *The Web of Life (1997)*:

> At each scale, under closer scrutiny, the nodes of the network reveal themselves as smaller networks. We tend to arrange these systems, all nesting within larger systems, in a hierarchical scheme by placing the larger systems above the smaller ones in pyramid fashion. But this is a human projection. In nature there is no 'above' or 'below', and there are no hierarchies. There are only networks nesting within other networks. (Id., p. 35)

Nothing in life is static. All is movement. The universe is a dance. In death processes, the relentless movement of life slows down and comes to a point of profound stillness. The unity of life was considered sacred by all ancient traditions, especially those of the East.

However, in this stillness is contained the grain for further movement, for new life. In every condition is contained its opposite. In stillness is contained movement, in movement is contained stillness, in hot is contained cold, in male is contained female. In the small boy is contained the great general, in the small girl is contained the famous film diva. In *yin* is contained *yang* and in *yang* is contained *yin*. What is contained is smaller as what bears it because it is in growth. By the same token, what bears the smaller is decreasing in size to become small itself. With culmination and fullness decay sets in, and a new cycle of growth is put in motion.

As holistic science moves on and becomes the reigning science paradigm within a few years from now, the distinction between science and philosophy will inevitably disappear. On the other hand, those who call themselves 'philosophers' and who are in fact nothing but gossipers and hair splitters, will equally disappear from the forum of so-called philosophy that gives them a warm spot now, and they shall be relegated to their petty web sites where they may talk the world into nothingness, messing *up all and everything into one gigantic soup of nonsense, calling this eternal mess, blasphemically so, philosophy* …

Science and Truth

The simple truth is that every science is *observation*. After observing nature, we recollect what we saw and share it with others, who in turn observed and reported. This is how any primal form of scientific consensus is brought about: by sharing observations. However, this has not been the normal course of our Western science. Instead of observing, nature was subjected under coercive scientific reason and this is how intellectual concepts or principles were *projected upon nature*. Every child understands that this is the best way to render a distorted picture of nature, but this is how it is and what we see today as what is arrogantly called science, a trash container of projections.

Quantum physics has given a majestic blow to this concept of lies, of a madhouse of pretension and human hubris in telling our Cartesian and reductionist scientists that they are part of the experiment, that their subjective and changing humanity cannot be discarded out of their pretendedly so clean laboratory experiments. With other words, as human nature is part of all nature, human nature is *entangled* with all that can be observed; hence, *there cannot be any objective kind of observation*, and there cannot be any objective science. Science is entangled with the unknown, which is the human. The simple truth is that if we kept true to our observations and simply, and honestly, shared them, we would create true science. But to do this, honesty, scientific and simple human honesty is required. This honesty is not part, and was never part, of the scientific establishment, while it may be present in individual scientists. True science, if this has ever existed, therefore is not established science, but at best the science that we call exploratory or experimental, or what is called alternative science.

And in this sense, in our society, children are the only true scientists, because they simply observe and report what they saw, without projecting any intellectual content upon their observations. And in this sense, all true scientists have a childlike way of doing science. Albert Einstein is the single best example for this truth.

Manly P. Hall, in his book *The Secret Teachings of All Ages (1928/2003)* affirms that in ancient traditions science and religion were not separated and that therefore science at that time was much closer to truth than it is today.

NATURAL ORDER

Manly P. Hall

Hippocrates, the famous Greek physician, during the fifth century before Christ, dissociated the healing art from the other sciences of the temple and thereby established a precedent for separateness. One of the consequences is the present widespread crass scientific materialism. The ancients realized the interdependence of the sciences. The moderns do not; and as a result, incomplete systems of learning are attempting to maintain isolated individualism. The obstacles which confront present-day scientific research are largely the result of prejudicial limitations imposed by those who are unwilling to accept that which transcends the concrete perceptions of the five primary human senses. (Id., p. 344)

Ervin László provides us with many honest statements of an avant-garde scientist, musician and genius that are unusual to hear from the mouth of a highly reputed scientist. According to László, science does not automatically equate truth, but sets up an equation of relationship between reality and scientific truth. This relationship is brought about and maintained through mapping reality with scientific instruments and theories. Ervin László explains in *Science and the Akashic Field (2004):*

> Science's disenchantment of the world has exacted a high price. When mind, consciousness, and meaning are seen as uniquely human phenomena, we humans—purposeful, valuing, feeling beings—find ourselves in a

universe devoid of the very qualities we ourselves possess. We are strangers in the world in which we have come to be. Our alienation from nature opens the way to the blind exploitation of everything around us. If we arrogate all mind to ourselves, said Gregory Bateson, we will see the world as mindless and therefore as not entitled to moral or ethical consideration. (Id., p. 14)

Whatever interpretation of the findings scientists may espouse, they are hard at work mapping ever more of the reality to which their observations and experiments are believed to refer. (Id., p. 16)

Whether or not scientific theories are humanly meaningful, they are clearly not eternal. Occasionally even the best-established theories break / down - the predictions flowing out of them are not matched by observations. In that case the observations are said to be 'anomalous'; they have no ready explanation. Strangely enough, this is the real engine of progress in science. (Id., pp. 16-17)

Investigating the anomalies that crop up in observation and experiment and coming up with the fables that could account for them make up the nuts and bolts of fundamental research in science. If the anomalies persist despite the best efforts of mainstream scientists, and if one or the other of the fables advanced by maverick investigators gives a simpler and more logical explanation, a critical mass of scientists (mostly young ones) stops standing by the old paradigm. We have a paradigm shift. A concept that was until then a fable is recognized as a valid scientific theory. (Id.)

But there is a danger of mapping reality with the tool of scientific investigation and relying exclusively on the worldview provided by science. Why has Western science never grasped the idea that life is basically energy, and why is it stuck in scientific materialism? Dr. Alberto Villoldo provides one of the answers, in his book *Shaman, Healer, Sage (2000)*. He says:

> Once we have drawn our maps of reality, 90 percent of our synaptic connections die. We become familiar with only one way to get to the river. The other routes are erased. (…) The spiritual landscape is not even acknowledged as real. There is no river, so why cut trails to get to it? Westerners have not developed the neural pathways to sense energy. (Id., p. 113)

Vidette Todaro-Franceschi, in her book *The Enigma of Energy (1991)*, has seen science and religion converging around the subject of her research: the nature and the enigma of energy. She writes in the introduction of her book:

> The more I worked on this project the more I became aware that somehow science and religion were converging. It was never my goal to merge these two seemingly disparate areas; in fact, when my search led me into religious realms of thought, I tried hard at first to stay clear of them. But it was impossible to do so. Anytime I came across literature that was related to an

idea of energy there were implicit or explicit spiritual overtones. Most surprising was the abundance of spiritual ideas found in physics. It seems that you simply cannot talk of wholeness or oneness without getting into some kind of religion.

When we agree that science is a process of mapping reality to conceptual perception to arrive at a relatively coherent view of the world, then we have to bring in as well the fact that this mapping of reality is a *consensus* after all. To view things realistically, we have to admit that if mapping reality was done only by one single scientist and if all other scientists disagreed, a new scientific paradigm would most probably not be formed.

So we can say that scientific observation of reality is something where the consensus of more than just one scientist is involved and required. It doesn't need to be all scientists, but what typically happens is that first a minority of scientists, a small yet powerful avant-garde group, forms a new paradigm which is then, gradually, taken over by a growing number of mainstream scientists. This is how scientifically approved reality is build, and rebuilt, in a constant process of renewed consensus. Now, Ervin László explains in *Science and the Akashic Field (2004)* that this process becomes more and more complex because the number of parameters involved in mapping reality to conceptual reality shoot up exponentially. This results forcefully in more and more 'esoteric' scientific models that need more and more research equipment, resources, and complex instruments to be verified over time:

> ERVIN LÁSZLÓ
>
> While conservative investigators insist that the only ideas that can be considered scientific are those published in established science journals and reproduced in standard textbooks, maverick researchers look for fundamentally new concepts, including some that were considered beyond the pale of their discipline but a few years ago. As a result, the world in a growing number of disciplines is turning more and more fabulous. It is furnished with dark matter, dark energy, and multidimensional spaces in cosmology, with particles that are instantly connected throughout space-time by deeper levels of reality in quantum physics, with living matter that exhibits the coherence of quanta in biology, and with space- and time-independent transpersonal connection in consciousness research—to mention but a few of the currently advanced 'fables.' (Id., p. 24)

Vidette Todaro-Franceschi, in *The Enigma of Energy (1991)*, writes that there are three periods in the development of the idea of nature, which she sees coincidentally reflect the ideas of energy.

THE NEW NATURAL ORDER

> In his discussion of the first period, the Greek view of nature, Collingwood points out that the ancient Greeks believed a certain vitality or ceaseless motion existed in nature, which they generally attributed to the soul. (…) The most important aspect of Aristotle's conception of nature lies in his belief that all things have a final cause, which is exhibited by the individual thing's form. According to him the soul was the essence of living things, and of course the form of anything / was the purpose or reason for its becoming. Overall, according to Aristotle, the teleological qualities of things were so strong that there could be no explanation for anything in nature, including us, without it. (Id., pp. 123-124)

In the second period, she reports, that Collingwood referred to as the Renaissance view of nature, mechanism was firmly established.

> Collingwood notes that the second stage of the Renaissance view of nature came about with the Copernican discovery that our world was not the center of the universe. The main contention during this time became 'the denial that the world of nature, the world studied by physical science, is an organism and the assertion that it is both devoid of intelligence and of life'. / During this period, human beings were seen as outside of, rather than a part of, nature. We became pompous, thinking that we controlled things and that we were somehow superior. Explicit in this view was the denial of final causation. The primary focus was on matter and the natural laws by which matter changes. Science and philosophy recognized only efficient causes: forces producing effects. And finally,

mathematical structure accounted for the changes, both of a qualitative and quantitative nature. (Id., pp. 124-125)

The third and last period identified by Collingwood, she reports, is the modern view of nature, which has its origin in the latter part of the eighteenth century when process and change became the focus.

> I believe that during this period the idea that energy was an autonomous existent contributed to the shift in focus. It became vaguely evident that change was inherent in various things; that is, it was recognized that change could occur without the provocation of external forces or efficient causes.
>
> In conjunction with this new take on an old idea that was present in both Eastern and Western antiquity is the increasing awareness that intuition plays a significant role in scientific discoveries. As the historical background of the idea of energy attests, intuitive ways of knowing have been crucial to the development of scientific ideas throughout history. Many individuals knew things, such as the energy conservation doctrine, without being able to empirically verify them. In other words, intuitive ways of knowing seem to have led / us in the right direction long before we were capable of scientifically validating what we somehow knew to be so.
>
> Subjectivity and subjective ways of knowing, such as intuition, have become as vital to our understanding as objectivity and empirical ways of knowing. In this modern view of nature humankind has once again come

to be recognized as being part of nature, rather than outside of it. (Id., pp. 125-126)

SCIENCE AND VIBRATION

Vibration really is the most basic of phenomena around life and living. Life is essentially vibration, and all vibration potentially is at the origin of one or the other form of life.

This was recognized by perennial philosophy, especially the Pythagorean branch of it. Vibrational and sound healing was the special knowledge of the sage, and this tradition that was perhaps first established by the Egyptians, became firmly rooted in Greek antiquity and from there went on till the Renaissance.

Vibration was considered as the Divine directly impacting upon matter, something related to the *cosmic breath*, and the knowledge about vibration was used for sound healing, healing with sound, with music. In addition, as noted by Manly P. Hall in *The Secret Teachings of All Ages (1928/2003)*, it was recognized that every element in nature has its own keynote. Hall writes:

> If these elements are combined in a composite structure the result is a chord that, if sounded, will disintegrate the compound into its integral parts. Likewise each individual has a keynote that, if sounded, will destroy him. The allegory of the walls of Jericho falling when the trumpets of Israel were sounded is undoubtedly intended to set forth the arcane significance of individual keynote or vibration. (Id., p. 256)

In the *Hermetic Theory Concerning the Causation of Disease*, seven causes and remedies were recognized. The second of these was vibration. Hall writes:

> The second method of healing was by vibration. The inharmonies of the bodies were neutralized by chanting spells and intoning the sacred names or by playing upon musical instruments and singing. Sometimes articles of various colors were exposed to the sight of the sick, for the ancients recognized, at least in part, the principle of color therapeutics, now in the process of rediscovery. (Id., p. 349)

And he leaves no doubt that vibrational medicine, as we call it today, was an established branch, if not the main branch, of healing:

> The magic rituals used by the Egyptian priests for the curing of disease were based upon a highly developed comprehension of the complex workings of the human mind and its reactions upon the physical constitution. The Egyptian and Brahmin worlds undoubtedly

understood the fundamental principle of vibrotherapeutics. (Id.)

Vidette Todaro-Franceschi asks in her fascinating scholarly study if modern scientific discoveries could be reconciled with Aristotle's idea of *energeia* as actuality?

She answers in the affirmative, explaining that science and philosophy 'are finally merging their beliefs as nature is being increasingly recognized as dynamic rather than mechanistic.' And what she writes here is of importance not only for the seeming dichotomy of science and philosophy, that is actually a mutually fertilizing and positively co-dependent relationship, but also for the vibrational nature of all creation.

> All things have a definite rhythmical pattern that is constantly changing. Activity that is probabilistic but not predictable is innate in all nature. Even on a quantum level things are predictably unpredictable! Movement is in a definite direction toward something not yet actualized. And the direction a thing moves in is for the benefit of its own becoming. Therefore, the modern view of nature and hence, our universe, can be equated with principles set forth by Aristotle centuries ago. (Id., p. 41)

NATURAL ORDER

The True Religio

Generalities

For a new natural order to emerge, it is essential that we relearn the true *religio*, which is the backlink to our true self, or selves. For this to happen, we need to get in touch with inside, and begin to dialogue with our inner selves, on a daily basis. This is what all native peoples do, when they are in trance, during festivities, or religious gatherings, or when they practice healing or self-healing through shamanic rituals.

All what happens in the world happens first of all inside of us, within our own inner landscape. That is why it is so tremendously important to begin all spiritual quest, and all journeying toward truth inside, in a state of quiet introspection. This knowledge is part of perennial philosophy.

Eric Berne, when creating *Transactional Analysis (TA)* in the 1950s, was not coming up with a novelty but with a scheme that *mapped insights* that the more wistful part of humanity had fostered since the beginnings of written history. As such, Eric Berne did a very important integrative work that has served

THE NEW NATURAL ORDER

healing and understanding of psychic processes, but that until this day never found its way to the rather imbecile and reductionist minds of the populace at large. However, this deep ignorance may well be not a lack of insight or learning, but the result of systematic manipulation and suppression of intuitive knowledge through the school system in all dominator cultures.

Our *inner selves* are energies in our psyche that form part of our total and integral wholeness. In the ideal case, they should be balanced and in harmony with each other.

This means that all inner selves should work as a sort of *inner team*. It is essential that all members of this inner team are fully awake and communicate with each other. In most people's psyches, however, it is as with that old mystic painting that depicts the inner child as a little angel who is somnolent or asleep. The worst condition of the inner child is the cataleptic inner child, the inner child that is deeply unconscious.

The Inner Selves

Eric Berne recognized three essential inner selves, *Inner Child, Inner Parent* and *Inner Adult*. In my own research and work with the inner dialogue during an Erickson hypnotherapy, I encountered the presence of additional entities such as the *Inner Controller* or *Inner Critic* as the instance in the psyche that represents the societal, cultural and moral values that we have internalized through education and conditioning. If the inner controller is hypertrophied and dominates the psyche, we are unable to realize our love desires.

In addition to these inner selves, I encountered an entity of superior wisdom that I called *Lux* (light) and a shadow entity I called *Sad King* and which embodied repressed *pedoemotions* that had turned into sadistic drives.

Inner Child

The inner child is an inner self, part-personality, or psychic energy, created between our 7th and 14th year of life, and that is part of our *inner triangle*. Positively, the inner child energy is primarily emotional and wistful, predominantly creative. It is the motor of every human being's creativity. Negatively, the inner child can be mute or cataleptic which means that its energy cannot manifest, or else its energy is turned upside-down which makes an inner child that is rebellious, capricious, willful or overbearing.

—See Peter Fritz Walter, Walter's Inner Child Coaching: A Guide for Your Inner Journey (2017).

Inner Adult

The inner adult is an inner self, part-personality or psychic energy that represents our logical thinking, our reason, our maturity. Positively, it makes for our balanced decisions, our down-to-earth attitudes and our sense for daily responsibilities. Negatively, the inner adult manifests as the intellectual nerd or through emotional frigidity, cynicism or an obsession to measure human relations on a scale of reasonableness or straightness without considering the emotional dimension. The hypertrophied inner adult energy plays a major role in modern education where it results in devastating damage on the next generations' emotional integrity.

Inner Parent

The inner parent is an inner self, part-personality or psychic energy that represents our inner value standards, our moral attitudes, our caring for self and others, but negatively also our judging others, our I-know-better attitude or blunt interference into the lives of others without regard for their self-reliance and privacy. The hypertrophied inner parent energy plays a dominant role in tyrannical and persecutory societal, religious and political systems.

Inner Dialogue

The inner dialogue is a technique conducive to getting in touch with our inner selves through relaxation or self-hypnosis and subsequent dialogues with one or several of our inner selves, in a state of light trance. This state of light trance can be self-induced, a technique that I demonstrate and explain in detail in my selfhelp audio books.

The inner dialogue should ideally be fixed on paper, at least in the beginning, because the voices that come up are very soft and writing down the dialogues helps to keep focus. The technique is also called *Voice Dialogue*, for example by Stone & Stone, in their *Voice Dialogue Manual*.

—Hal and Sidra Stone, Embracing Ourselves: The Voice Dialogue Manual (1989).

However, the expression could mislead novice users as the 'voices' are not really voices, as they are not to be heard with our ears, but something like *intuitions, or flashes of intuition,* or sudden precisely formulated thoughts that seem to come 'from nowhere'.

Multidimensionality of the Psyche

Every healthy psyche is composed of a multitude of energies or entities, and that it is through our ego that these entities are working under a certain roof structure of conscious control. Otherwise, if this ego, for whatever reason, disappears, we enter the realm of schizophrenia, which can be, as in psychedelic trips, a welcome temporary condition, or a long-lasting psychosomatic illness.

Function of the Ego

The function of the ego is not to dominate any of our inner entities, but to *orchestrate* them, to direct them in a team-like cooperation, such as for example the conductor of an orchestra leads more than one hundred musicians to play in sync in order to reproduce a musical score with accurate precision and harmonious sound. This is the function of the healthy ego within the multidimensional psyche. Needless to add that with most people the ego and the inner controller are hypertrophied and dominate if not suppress all the other inner entities which is the explanation for why such a high percentage of the world's population is completely uncreative, dull and imitative in their behavior, and why they use only about five to eight percent of their emotional and creative intelligence potential.

Toward a Science of Life

Emotional Flow

Emotional Flow is a notion I have developed in the context of my research on the bioenergetic etiology of *sexual paraphilias*, and it describes the natural flow condition of our emotions, when no distortion has taken place in the bioplasmatic setup through an alienating moralistic education, and/or the suffering of emotional abuse in the form of an ongoing parent-child co-dependence in childhood and/or adolescence. It could as well be called *emotional sanity*, for this is what we are talking about here.

THE NEW NATURAL ORDER

My research on sexual paraphilias provided conclusive evidence for the fact that every human being possesses a unique *emotional identity code*, something like a vibrational ID number, that works like a cosmic identifier and sets us apart as absolutely unique beings. This is valid not only for humans but this vibrational pattern is unique also for animals, for plants and even for inanimate matter such as rocks.

The Nature of Emotions

While today's mainstream psychology to some extent admits the cognitive nature of emotions, it relates emotions to thought and perception only and locates them in the brain, while the overwhelming number of perennial science traditions and newest research on the human energy field shows that *emotions are located in the human aura* and possess an *inherent quality of flow*, as well as their own *intrinsic intelligence*. Thought and emotions are *vibrations* that flow through our etheric or luminous body. This was recognized as early as in the 1970s by Valerie Hunt, and was summarized in her book *Infinite Mind (2000)*. In this sense, also animals and plants have emotions, which was something completely discarded or overlooked by traditional psychology, while Wilhelm Reich, as early as in the 1930s, was on the right track in his *bioelectric evaluation of emotions*, writing that emotions are specific energy functions of the protoplasm.

Emotions are manifestations of the *life force* in the living organism. They are to be found as biogenetic and bioenergetic vibrations in the cell plasma. In this sense, emotions are functional, and they are directly related to all visceral life functions. An indication that an organism has died is the absence of emotional flow. This is valid also for human beings and in the sense that while people may physically be well alive, they may be emotionally dead since many years.

> WILHELM REICH
>
> I stress the rationality of the primary emotions of all living. The mechanists of depth psychology have namely spread the view that all emotions were but drives and therefore irrational. However, emotions are specific functions of the protoplasm. Emotions and the natural movement of the bioplasm are functionally identical phenomena.
>
> —Wilhelm Reich, Äther, Gott und Teufel (1983), p. 54. (Translation mine).

The *mainstream* scientific view on emotions is *reductionist, ignorant, based on denial*. For example, mainstream psychology considers emotions to be cognitive elements, thus elements of thinking. More precisely, Western science denies emotions their intrinsic energy or wave character, thus putting them on one level with matter. By contrast, my research shows that emotions are energy, pure energy. As a result of the reigning reductionist view, mainstream research on emotions uses statistics and looks at human behavior, instead of looking at emotions as *energy manifestations*. With one word, Western mainstream science is completely in the dark about the true nature of emotions.

The *alternative scientific view* recognizes that emotions are actively involved in the energy metabolism of the organism, that they are kaleidoscopic in nature and functional, and that they serve cognition through emotional intelligence.

—See, for example, Candace B. Pert, Molecules of Emotion (2003).

The *spiritual view* of emotions, as it is part of perennial philosophy and most science traditions of the Middle East and the Far East can be summarized with the formula 'Emotions are God.'

Natural Order

Emotional Awareness

The conscious perception of our emotional flow includes awareness of our emotional predilection and sexual attraction in every given moment or situation. For example, a nurse should be conscious of her emotional flow regarding patients she is working with, and an educator needs to develop *emotional awareness* regarding their natural *pedoemotions* toward the children they are working with.

The bioenergetic current that flows through the organism, from the cell plasma to the periphery and into the luminous body and again back from the luminous body to the cell, depends on the polarity of the current. When it is positive, it is expansive and flows from the cell to the periphery (joy), when it is negative, it retreats from the periphery back into the cell (fear).

These principles of flow that are inherent in the nature of the bioenergy are also to be applied in the etiology of sexual and non-sexual sadism. In the natural sexual streaming of the bioenergy, that Wilhelm Reich described as 'hot, melting emotions,' the energy during orgasm explodes from the cell toward the luminous body. In sadism, however, because of the muscular armor in the pelvis region and other parts of the body, the energy cannot freely flow outwardly and therefore is repelled back with the result that instead of relaxing joy and expansive feelings, what is felt after orgasm is depression, anxiety, and fatigue.

These latter symptoms can also be used as signals for diagnosing sadism. Hence, it is actually possible *to heal sadism by getting the emotional current to flow again naturally through he organism.* This can be done through muscular relaxation or through work on consciousness, using *Life Authoring*, or else a combination of these with body work, massage or martial arts techniques.

> —Life Authoring, or Author Your Life is an awareness building self-coaching technique I have developed. Detailed information about this method is contained in my three books on coaching and training, Walter's Leadership Guide (2017), Walter's Inner Child Coaching (2017), and Walter's Career Guide.

Emotional Balance

Children and babies naturally, when they are swinging in their continuum balance, are within the realm of emotional integrity. Emotional sanity is manifest when emotional energy is integrated, which is the natural condition in the living organism. This blessed condition can also be called *emotional balance*. Integration occurs ideally on three levels:

- Multisensorial (Spirituality)
- Extrasensorial (Parapsychology)
- Sensorial (Eroticism, Sexuality)

To bring about and maintain *emotional sanity in relationships with children* is the task of every parent and every educator; it means to care for the natural continuum balance of the child to remain untouched and preserved.

This requires in practice to observe a principle of *sacred non-interference* in the child's continuum, to restrain from inflicting educational violence on the child, to respect the child's privacy, to respect the child's social life, which means to *abstain from controlling the child's relationships/friendships* with peers and other adults, to give the child real opportunities for love and sexual relations outside of the family, to restrain from emotional and sexual incest with the child, and to help the child accept their body and their emotions through loving dialogue about all matters, without taboo.

This educational task also means to preserve that child's natural bioenergetic setup from birth, the free flow of the vital energies in their organism, the healthy vibrancy of the aura and bioplasma, the natural cycle of charge and discharge during sexual orgasm, and this for the whole life cycle of the person, from conception to death. Our emotional setup is by nature harmonious and self-regulated, and it favors equitable relationships, love and natural sharing of emotions, joy, and goodness. It becomes distorted through early interference with the natural energy pattern in form of educational violence and abuse, and the obstruction of the emotional flow through the educational prohibition of expressing emotions and sexual wishes.

Emotional Intelligence

Emotional intelligence is one of the four types of intelligence, which are logical-rational intelligence, emotional intelligence, graphical-spacial intelligence and tactile intelligence. Emotional intelligence is especially active when it goes to understand relationships, human affairs, and the psychological implications within them. Age-old wisdom that asserts that women are more emotionally intelligent than men was corroborated by modern experiments, for a majority of subjects. However, it has to be seen that many highly intelligent men are also highly emotionally intelligent.

—See, for example, Daniel Goleman, Emotional Intelligence (1995).

The perhaps most typical example is Albert Einstein who, despite being a brilliant logical thinker, was also highly emotionally intelligent, artistic, and creative, a musician and a fine psychologist, next to being one of the greatest physicists and mathematicians who ever lived on the globe.

—See, for example, Joyce Goldenstein, Albert Einstein: Physicist and Genius (1995).

The Life Force

What native cultures tend to call the *life force* was termed very differently over the course of human scientific history. Here are some examples:

- Cosmic Life Energy

THE NEW NATURAL ORDER

- Bioenergy
- Élan Vital
- Vis Vitalis
- Spirit Energy
- Vital Energy
- Cosmic Energy
- The Field
- Zero-Point Field
- A-Field
- L-Field
- Akashic Field
- Human Energy Field
- Ch'i
- Ki
- Mana
- Prana

- Wakonda
- Hado
- etc.

The Emonics Terminology

This list is not exhaustive. I have tried to fill in what I could to make this new vocabulary on the life force as comprehensive as possible, but it still is only a sketch. Much further research is needed.

Akashic Records

Akashic Records are the universal memory consisting of energy patterns. It is a cosmic memory where all and everything is stored including a cosmic library of archetypes. The library is accessible to psychics.

Aura

The aura of living beings (plants, animals, humans) is an expression to designate the energy body, ethereal body or energy spheres around the physical body. The total number of these spheres is 7.

THE NEW NATURAL ORDER

E and E-Force

E is the functional complement of consciousness. Consciousness is a function of E. E and consciousness are a functional whole. E manifests on this planet as e-force. Shifts in consciousness bring about shifts in e-force. Shifts in e-force trigger states of altered consciousness. Superconsciousness is a state of e-force at its peak level.

Emonic and Demonic

Disintegration of emotions occurs through denial and repression. The results are violence, regression and sadism—which are obstacles to evolution.

Emonic Charge

Emonic charge is the biogenic positive charge accumulated in living organisms that leads, typically, to discharge in the form of ecstatic convulsions or sexual orgasm and that is part of the inherent self-regulatory system of the cell.

Emonic Awareness or Emotional Awareness

Emonic awareness is the conscious perception of our emonic flow, which includes conscious perception of our emotional predilection and sexual attraction in every given moment or situation. For example, a nurse should be conscious of her emonic flow regarding patients they are working with, or educators need to develop emonic consciousness regarding their natural pedoemotions toward the children they are working with.

Emonic Current, Emonic Flow or Emotional Flow

Emonic flow is the bioenergetic current that flows through the organism, from the cell plasma to the periphery and into the luminous body and again back from the luminous body to the cell, functionally related to the polarity of the current. When it is positive, it is expansive and flows from the cell to the periphery (joy), when it is negative, it retreats from the periphery back into the cell (fear). Emonic Flow, in popular language, may be expressed simply as *emotional flow*.

Emonic Integrity

Children and babies naturally, when they are swinging in their continuum balance, are within the realm of emonic integrity.

Emonic Setup

Emonic setup is our natural bioenergetic setup from birth, the free flow of the vital energies in our organism, the healthy vibration of our aura and bioplasm, the natural cycle of charge and discharge during sexuality, during the whole life cycle from conception to death. Our emonic setup is by nature harmonious and self-regulated, and it favors equitable relationships, love and natural sharing of emotions, joy, and goodness.

It becomes distorted through early interference with the natural energy pattern in form of educational violence and emotional abuse, and the obstruction of the emotional flow through the educational prohibition of expressing emotions and sexual wishes, typically when the educational paradigm is based upon moralism.

Emonic Vibration

Emonic vibration is the bioenergetic flow and unique vibrational code inherent in every living organism, without which life would cease and death would follow. Emonic vibration is thus an immediate characteristic of life.

NATURAL ORDER

Corroboration of Emonics

The ancient Vedas teach exactly that, but in a metaphorical language. So does Yoga, Qigong, Zen, and other esoteric spiritual practices.

As noted by Manly P. Hall in *The Secret Teachings of All Ages (1928/2003)* at page 256, it was recognized that every element in nature has its own keynote.

> If these elements are combined in a composite structure the result is a chord that, if sounded, will disintegrate the compound into its integral parts. Likewise each individual has a keynote that, if sounded, will destroy him. The allegory of the walls of Jericho falling when the trumpets of Israel were sounded is undoubtedly intended to set forth the arcane significance of individual keynote or vibration.

Emonics and Sexual Paraphilias

I have created Emonics for achieving scientific recognition and validation of the fact that all sexual attraction is based upon *emotional predilection*, and not the other way around; sexual attraction is the result, and not the cause, of emotional attraction and predilection.

My theory is thus opposed to modern sexology, which assumes that sexual drives are independent of emotions, that they have a life of their own, and that they are somehow robotic and mechanical. Sexology, as it is practiced today, is mechanical, not the human being that it scientifically scrutinizes and examines.

I got to coin the term *emonic* as the logical counter-value to the term *demonic*, and all started from there. I was asking myself namely how it could be that our language has well coined the expression demonic, but *not* the expression emonic, as its natural positive opposite? The reason is simply that Occidental science has never recognized nor integrated the bioenergy, *life force*, or cosmic life energy.

What I did thus was to retrace the alternative science tradition that I found has a perennial history both in the East and the West, while it was almost always underground. And I found what I was searching for, there is a holistic science tradition to be retraced back to times immemorial that has integrated and described in detail the cosmic life energy and how all life benefits from it, or, in case of the human, acts counter to it. This science was however never *unified* and it lacked a *common code, a vocabulary*; thus I simply created that vocabulary, and that was it.

Thus, the term *Emonics* describes not only the fact that emotions and sexuality are in a continuous swing, but it also can explain how of a loving erotic embrace can emerge from what we use to call love, as something that represents a basic unity.

Emonics and Emosexuality

From this point of departure, *Emonics* describes the complex process of interrelations between our emotions and our sexual attractions that we experience in the moment we begin to love another person. The term 'sexuality' has in fact very little significance because it limits itself at genital activity. I therefore introduced the term *Emosexuality*, which is much larger a term allowing *sexology* and *cognitive psychology* to better observe and describe the results of their research on the significance of human love and love relations between humans, and also, exceptionally, between humans and animals.

Emonic Dysfunctions

Emonic dysfunctions are pathologies either on the emotional level or on the sexual level or on both levels at the same time that negatively affect, or even render impossible, the natural and plain erotic satisfaction during the loving embrace.

One important task of Emonics is *prevention*, and healing of emonic dysfunctions. An important research result is the fact that when emotional and sexual streaming is blocked, what results is *demonic* emotions, which are sadistic, dominating, negative and destructive. It is for this reason that *Emonics* has an important role to play in peace research.

Primary Power and Live Your Love

Primary Power, Soul Power, Self Power

The new natural order can only be founded upon *permissiveness* and *self-regulation*, which is why it must abandon moralism and morality-based sex laws. This implies a return to the state of *primary power*, the power that is a natural part of autonomy. Contrary to worldly power, primary power or soul power is not a pent-up power urge for debasing and dominating others, but a state of inner harmony that seeks harmony and good relationships with others.

NATURAL ORDER

Soul power, primary power or *self-power* can be defined as the natural and non-abusive power of a basically sane human being. *Primary power* is the natural power that the sane child develops when allowed to grow into autonomy and self-reliance. This is namely the case when the postnatal primary symbiosis between mother and child during the first eighteen months of the newborn was a positive experience for both mother and child, and when the mother can allow the infant to gradually grow into autonomy as the child widens his or her grasp and perception of the environment and thus gradually leaves the condition of primary narcissism.

Soul Power, which I synonymously call *Primary Power* or *Self Power* is a concept I created to connote and describe our original power; this innate power is based upon *innocence*. It is distinct from the harmful secondary or worldly powers, which are based upon knowledge, that profoundly mark our current society, and which are clearly violence-inducing, and in the long run damaging the human potential and natural human spirituality.

Developing soul power is conceptually linked to developing awareness of our intrinsic *soul values* that typically, and in the regular case, do not coincide with our accepted social values. So there is at the starting point an inner conflict, or duality, between our soul values and our cherished and agreed-upon social values. This inner conflict must not be silenced, but met with passive (and peaceful) awareness for this inner conflict is actually creative and brings about soul power in a way completely different from what fashionable life coaches such as Anthony Robbins are teaching and practicing, and what they use to call *personal power*.

—See, for example, Anthony Robbins, Awaken the Giant Within (1991).

By contrast, I denote as *secondary powers* the largely *abusive powers* that result from the fragmented, schizoid and overtly narcissistic mainstream individual that incarnates the core personality of *Oedipal Culture*.

Different authors and scientists have given this dichotomy different names. Sigmund Freud and after him psychoanalysis as a whole speak of *secondary drives*, where I use the expression *secondary power*. Gary Zukav, author of *The Dancing Wu Li Masters (2001)*, speaks of authentic power vs. external power.

NATURAL ORDER

LIVE YOUR LOVE

Live Your Love is an empowerment concept I created back in 1998. It was the fruit of years of work on soul power, soul values, and on my self-help productions and consciousness guides.

In a society that is highly judgmental and labeling as our postmodern international consumer culture, it is of paramount importance to define all the ingredients that identify you, that build your own personal and unique identity. The most important ingredients in this soup of yourself are your soul values, and your *love*.

Sex research has amply demonstrated that there is no single other human behavior than sex that is to that point diversified and different from one person to the next. You can say there are no two humans on the globe who have exactly the same sexual attraction and whose sexual appetite and table manners are even similar, let alone identical. That means when you are sexual you are *you*, when you refuse to be sexual (because of fear of being different), you are nobody.

Live Your Love means to incarnate your loving attraction in this dimension in the form of precise tastes and dishes to consume. My advice is that if big brother tells you that your taste is perverse, just grin and continue eating … ! And think of one important thing, conclusion of all criminological research: 'All sexual monsters are virgins or very inexperienced lovers!'

When you live your love and love your life, you are on the good side of the river.

Permissive Education

A return to the natural order is only possible if we can form collective agreement for *drafting a truly humanistic, emotionally sane and permissive education* that is focused upon catalyzing the natural gifts and talents of the child, and that may turn ultimately toward the creative child.

In the past, permissive education was felt to go against the stream of patriarchal society and monotheistic religion. However, despite the consensus for a repressive approach to education, authors of high distinction provided contributions to improving the lot of our children.

NATURAL ORDER

Some of these men and women were John Locke, Thomas Hobbes and Jean-Jacques Rousseau, Wilhelm Reich, Alexander S. Neill, J. Krishnamurti, Alice Miller, and others. It has to be noted, however, that these honorable authors, except Reich, Neill and Miller, never talked about children's emotional and sexual needs.

In the meantime, modern pediatrics and psychology have understood that permissive education is a must in a truly democratic society. Back in 1973, however, sexual permissiveness was still declared a myth in Western society.

—See John P. Alston, Francis Tucker, The Myth of Sexual Permissiveness (1973).

This is quite astonishing as this research was published at a time when social values were a lot more permissive compared to today.

Introduction

For anyone who wishes to know about permissive education in a culture that hates the child, there is no way other than reading literature. In the great literature of all times, I found individual parent-child relations described that did not fit in the normative scheme and where the parent was sensitive enough to give the child headroom for autonomy and non-regulated intimacy.

It is certainly good to read and study the above mentioned educational projects, but as Alexander Lowen once wrote to me as a reply to a letter I had sent to him about my own educational project, every school can only be as good as the educators who run them. There is no education on paper. All in this field needs to be humanized and made fit in the daily little critter of relationships. I have faced the worst educators in persons who are *high-strung idealistic* and have a lot of theory in their heads, and the best in those who are *simple-minded, but attentive to detail*, fresh, loving and spontaneous.

This brings me to talk about the character structure that fosters permissiveness. It goes without saying that it's a character that is neither neurotic nor sadistic, but *genital* in the Freudian sense or *orgasmic* in Reichian terms. In my own terms, I would say it's a person whose *emonic flow* is intact and where *desire is conscious*, or has been rendered conscious through building emonic awareness.

In our society, as long as things are emonically as they are, there is no hope to expect a change toward permissive education, as the character structure of those in power and those in power in education is sadistic because they have *repressed* their *pedoemotions*. And without an opening here prior to any change, there will be no change. This is how it is.

The Failure of Moralistic Education

Summerhill was founded in 1921 in a village near London, England. It was a free school which means that there was no moralistic education and no punishments.

What do I mean when I say *moralistic* education? While many people think that also Montessori and Steiner schools practiced an education that is free of repression and moralistic concepts, I found by observation that this is not true. Montessori looks very pragmatic, very rational and focused upon the necessities of daily life. Children learn to iron shirts, to do gardening, to cook. They are put in sensitization classes to stimulate their sensorial perception. However, this rhetoric is deeply false. The child does not need to be stimulated sensually since it is naturally sensual.

But of course, the moralistic and in this case Catholic background of Montessori education does not permit the children to live out their emotions and primarily their sensual and tactile needs. So they are first emotionally starved and then artificially induced into fake-feelings in so-called sensitization classes – a ridiculous idea altogether.

With Steiner, it's the stressing of the soul values of the child, as if the child by itself was not able to realize their soul values. Behind these different approaches are different basic philosophies regarding the role of the child in society.

Neill was against Montessori that he considered as a milder and more intellectual but nonetheless *intrinsically authoritarian* form of modern education. There is no doubt that Maria Montessori who was a believing and practicing Christian wanted to revolutionize education and her contribution for more humanity and respect for children was certainly authentic. Her teachings brought about amazing change, not only in her own schools. One of her inventions was the child-oriented seating furniture that we now all know from modern day care centers. But was it really an approach that served children to be better able to realize their own intrinsic nature?

Montessori's point of departure was the observation that the child's brain, not unlike a sponge, absorbs the intrinsic atmosphere of his or her environment. In her book *The Absorbent Mind (1973/1995)*, she cites psychological research that proves that children learn in the first three years of their life more than adults in sixty years of hard study. This is certainly true.

Although Montessori was in the beginning against all educational programs, she designed a specific educational curriculum for her schools that focused primarily on the intellectual training of the child.

By means of a sophisticated system of different games, puzzles and assembler games (that are much more complicated than what usually can be bought in toy stores), the child's mind is well prepared to handle all such equipment with competence and care, accidents being a rare exception.

My experience of this approach through visiting Montessori schools was rather negative. First it was extremely difficult to get a permission for visit at all. I had to justify my wish in a way as if I wanted to visit a secret terrain of the armed forces. The permission was conditioned upon my being very short and my restraining from any communication with the children. The children I saw seemed to be robots, pale, dull, insensitive, without life. But they worked, and how! Their way to work through the various tasks seemed obsessional, almost neurotic, while they were bombarded with a full-cry Beethoven symphony from a portable stereo.

These schools were places without soul and without humor. When the children had their pause, they sat down immediately on a bench, opened their lunch boxes and eat silently, without talking to each other. They seemed to have no contact to each other, isolated in their intellectual cages, without all what usually makes the somewhat noisy and happy society of natural carefree children. These children were little adults that had lost their souls and stared in the air with moody faces.

I felt highly irritated when I left those places. I found them even worse than the violent orphanages I had seen in some third-world countries. This was child dressage of its best, but one that was no circus because here not even clowns were allowed.

I never got a chance to visit Summerhill. I doubt that it is as permissive as it pretends to be, and this simply because it is located in England.

Can one imagine a more repressive culture as to the expression of natural human emotions? Of course, it is much more humane than most traditional schools and, in particular, corporal punishment is absolutely taboo. The education is permissive regarding the healthy emotional and sexual development of the child. Masturbation is not repressed, sexual play only for the purpose to avoid procreation. That is what we learn from Neill's books, but it must be doubted if in practice, in Puritan England, the free sexuality of children and youth is tolerated in a school.

Nonetheless, Summerhill has followed up to a great tradition. Neill's educational approach can be seen on a line with famous historical educational methods, like Jean-Jacques Rousseau's or John Locke's, in that it was founded upon the view that nature is generally good.

THE NEW NATURAL ORDER

—See Jean-Jacques Rousseau, Émile ou de l'éducation (1762/1964) and John Locke, Some Thoughts Concerning Education (1690/1823).

Summerhill is thus in flagrant contradiction with Christian, and in particular Calvinistic, child-rearing which uses harsh punishments and emotional frustrations 'to better the human soul' that it considers as basically bad and rotten. The goal of Summerhill is not to bring about conformists, but adults with high self-esteem, strong intuition and sensitivity, humor and respect for life in all its forms.

—Alexander S. Neill, Summerhill (1961), pp. 29 ff. and Neill! Neill! Orange Peel! (1972).

One of the main motivations Neill had for his school was to produce adults with a generally positive and constructive mindset, people who are free of hatred and the repressed anger that is part of traditional education; people also whose emotional life is intact and lively.

When Neill opened Summerhill, he was already fifty-one years old. He had spent years with studying human history and education and was deeply concerned that human history was marked by hate and violence, human destructiveness, intolerance, war and slavery. Neill got to see behind the veil of lies of modern civilization. He saw this pervading hatred again and again in the children who arrived from traditional institutions where they had been declared *uneducable*; he saw that the whole concept of the 'difficult child' was a myth in that these children were *not more destructive than others*, but unhappy, lonesome, emotionally blocked, and frustrated. Their destructivity and violence was but a symptom of the underlying reasons that were deeply rooted in the hypocrite, paranoid and violent societal system they were raised in.

In addition, most of them had been neglected or even abused by their parents or by educators in the homes they were coming from.

These children had lost trust in adults. They had been deceived and felt the cold pressure that comes from authoritarian ways of child-rearing. They did not know what love was about and being loved. In this sense, Neill acknowledged, all children who are raised in an authoritarian and repressive system will be 'difficult' once they are freed from the pressure they are subjected to. This difficult behavior, Neill found, actually was an inner healing process that established a new value system in their mindset. But first they explode, of course. With adults it is the same, as we all know. Violent crimes, war, slavery, torture and terrorism are the results of the hypocrite make-believe that we call civilization and that has in truth nothing to do with being civilized.

These consequences of the authoritarian educational system show that this approach is based upon wrong premises, and that it is not human and not made for humans since it *disregards human dignity* in the most flagrant way.

Neill knew that only an education that is based upon love can finally overcome the violence inherent in a society that is full of hatred. For the only way out of violence is fighting its roots: lack of love and respect, lack of positive encouragement, dehumanizing treatments and a belief system that is based on the idiotic and arrogant idea that nature is fundamentally bad and has to be improved.

Referring to Wilhelm Reich's research, Neill found that a moralistic education does not only negatively impact upon the mindset of children, but also infringes upon the soma, especially the *emotional and muscular balance* in the body of the child.

NATURAL ORDER

—Alexander S. Neill, Summerhill (1961), p. 207.

Also in accordance with Wilhelm Reich, Neill applied in his school the principle of *self-regulation*. Every attempt to impact upon the children in an intellectual or *moralistic* way was leading to failure and was soon abandoned. Instead, Neill believed that children comply with what they really subscribe to and understand so that they do it because they believe that it has to be done.

This attitude of course requires that the educator knows and believes that children are beings born with reason and that they will use this reason whatever their age is.

That is exactly the point where moralistically oriented educators have their deepest doubts. They dig a ravine between adults and children, as if children lived in another world with different natural laws. While they acknowledge the necessity of reason, they deny that children possess reason and rather put children on one level with animals.

There are passages in Calvinist religious pamphlets that are very explicit in this regard.

THE NEW NATURAL ORDER

Raising Humane Humans

Free child-rearing is unthinkable without conceding children their natural erotic and sexual feelings and, even more importantly, the freedom of speech regarding these feelings.

> —See Floyd M. Martinson, The Sex Education of Young Children (1981), 51 ff.

The latter is as important as the former because children who are allowed to do it without being allowed to talk about what they have done will never really believe that this freedom has been given to them. Instead they will experience guilt and shame.

Only a child who is sexually free and erotically satiated will develop his full potential of interest and work energy. It is further necessary that children really feel accepted as persons in their own right, in their natural wholeness that encompasses a virgin *emonic* setup that wants to be developed and experienced.

All life asks for growth, and it is therefore total nonsense when mainstream psychologists pretend the child had to grow with a sleeping sexuality that would suddenly awaken at the end of puberty. Fairy tales.

NATURAL ORDER

We are facing this challenge for a more truthful education also on a collective and global level in that we gradually develop more tolerance, understanding and compassion for others. This global change is brought about through changing the basic foundation of our educational and pedagogical values, and here I am talking about *public sanity*.

The Summerhill concept realized a first step toward more humanity here on earth, by raising more humane humans. Interestingly, the overwhelming majority of Summerhill graduates were later seen to score very well on the social ladder while *leading healthy and balanced lives* and experiencing positive and highly rewarding relationships. Neill stated that his own measure of success was the capability to work with joy and to live positively.

—Alexander S. Neill, Summerhill (1961), p. 29.

After forty years of experience with Summerhill, Neill was able to conclude that, applying this definition of success, 'most of the Summerhill graduates became successful people.' (Id.)

POSTFACE

Where Are We Now?

When we see that Summerhill started out in the 1920s, we should think that now, almost a century later, we should be with two feet in the new age—but are we? It seems to me we rather retrograded far behind the 20th century in all the basic areas of social policy making.

Only to look at the international 'sex offender' witchhunts, I feel we are right back in the dark ages, with the difference only that the church was replaced by the modern police state and the priest by the psychiatrist.

In other words, I would localize us right now at the end of the phase of antithesis. This end phase can extend over the next twenty or so years, but it may also be half a century until the synthesis is reached.

Of course, this varies from one society to another. Some progressive European societies such as Norway or nowadays, Spain, are already implementing social policies that clearly indicate they are 'thought' from the synthesis perspective.

For a nation such as the United States, however, where things look particularly dim in this respect, a social philosopher needs to be mentioned who clearly prepared the ground for the synthesis perspective. It was Joel Feinberg.

— See, for example, Joel Feinberg, Harmless Wrongdoing: The Moral Limits of the Criminal Law, Vol. 4 (1990).

It seems that his appeal for the government to avoid criminalizing natural and harmless human behavior was not really influential for the very contrary is being done over the last two decades or so.

I have worked out a comprehensive new agenda for social policy making in my book *The 12 Angular Points of Social Justice and Peace: Social Policy for the 21st Century, 2015/2017*, which is 'thought' exclusively from the synthesis perspective and which goes beyond Feinstein's suggestions in that it attacks the recent exorbitant getting-tough policies regarding 'sex offenders,' even those who are barely older than their victims and even of equal age, called 'juvenile sex offenders.'

Bibliography

ABRAMS, JEREMIAH (ED.)

RECLAIMING THE INNER CHILD
New York: Tarcher/Putnam, 1990

ALSTON, JOHN P. / TUCKER, FRANCIS

THE MYTH OF SEXUAL PERMISSIVENESS
The Journal of Sex Research, 9/1 (1973)

APPLETON, MATTHEW

A FREE RANGE CHILDHOOD
Self-Regulation at Summerhill School
Foundation for Educational Renewal, 2000

ARCAS, GÉRALD, DR

GUÉRIR LE CORPS PAR L'HYPNOSE ET L'AUTO-HYPNOSE
Paris: Sand, 1997

ARIÈS, PHILIPPE

L'ENFANT ET LA FAMILLE SOUS L'ANCIEN RÉGIME
Paris, Seuil, 1975

CENTURIES OF CHILDHOOD
New York: Vintage Books, 1962

GESCHICHTE DER KINDHEIT
Frankfurt/M: DTV, 1998

ARNTZ, WILLIAM & CHASSE, BETSY

WHAT THE BLEEP DO WE KNOW
20th Century Fox, 2005 (DVD)

DOWN THE RABBIT HOLE QUANTUM EDITION
20th Century Fox, 2006 (3 DVD Set)

RELATIONSHIPS AND LIFE CYCLES
Astrological Patterns of Personal Experience
Sebastopol, CA: CRCS Publications, 1993

ATLEE, TOM

THE TAO OF DEMOCRACY
Using Co-Intelligence to Create a World That Works for All
North Charleston, SC: Imprint Books / WorldWorks Press, 2003

BACHELARD, GASTON

THE POETICS OF REVERIE
Translated by Daniel Russell
Boston: Beacon Press, 1971

BIBLIOGRAPHY

Baggins, David Sadofsky

Drug Hate and the Corruption of American Justice
Santa Barbara: Praeger, 1998

Bagley, Christopher

Child Abusers
Research and Treatment
New York: Universal Publishers, 2003

Balter, Michael

The Goddess and the Bull
Catalhoyuk, An Archaeological Journey
to the Dawn of Civilization
New York: Free Press, 2006

Bandler, Richard

Get the Life You Want
The Secrets to Quick and Lasting Life Change
With Neuro-Linguistic Programming
Deerfield Beach, Fl: HCI, 2008

Barbaree, Howard E. & Marshall, William L. (Eds.)

The Juvenile Sex Offender
Second Edition
New York: Guilford Press, 2008

BARRON, FRANK X., MONTUORI, ET AL. (EDS.)

CREATORS ON CREATING
Awakening and Cultivating the Imaginative Mind
(New Consciousness Reader)
New York: P. Tarcher/Putnam, 1997

BATESON, GREGORY

STEPS TO AN ECOLOGY OF MIND
Chicago: University of Chicago Press, 2000
Originally published in 1972

BENDER LAURETTA & BLAU, ABRAM

THE REACTION OF CHILDREN TO SEXUAL RELATIONS WITH ADULTS
American J. Orthopsychiatry 7 (1937), 500-518

BERNARD, FRITS

PAEDOPHILIA
A Factual Report
Amsterdam: Enclave, 1985

BERTALANFFY, LUDWIG VON

GENERAL SYSTEMS THEORY
Foundations, Development, Applications
New York: George Brazilier Publishing, 1976

BIBLIOGRAPHY

BESANT, ANNIE

AN AUTOBIOGRAPHY
New Delhi: Penguin Books, 2005
Originally published in 1893

BETTELHEIM, BRUNO

A GOOD ENOUGH PARENT
New York: A. Knopf, 1987

THE USES OF ENCHANTMENT
New York: Vintage Books, 1989

BOHM, DAVID

WHOLENESS AND THE IMPLICATE ORDER
London: Routledge, 2002

THOUGHT AS A SYSTEM
London: Routledge, 1994

QUANTUM THEORY
London: Dover Publications, 1989

BOLDT, LAURENCE G.

ZEN AND THE ART OF MAKING A LIVING
A Practical Guide to Creative Career Design
New York: Penguin Arkana, 1993

HOW TO FIND THE WORK YOU LOVE
New York: Penguin Arkana, 1996

ZEN SOUP
Tasty Morsels of Zen Wisdom From Great Minds East & West
New York: Penguin Arkana, 1997

THE TAO OF ABUNDANCE
Eight Ancient Principles For Abundant Living
New York: Penguin Arkana, 1999

BORDEAUX-SZEKELY, EDMOND

TEACHING OF THE ESSENES FROM ENOCH TO THE DEAD
Sea Scrolls
Beekman Publishing, 1992

GOSPEL OF THE ESSENES
The Unknown Books of the Essenes
& Lost Scrolls of the Essene Brotherhood
Beekman Publishing, 1988

GOSPEL OF PEACE OF JESUS CHRIST
Beekman Publishing, 1994

GOSPEL OF PEACE, 2D VOL.
I B S International Publishers

BRANDEN, NATHANIEL

HOW TO RAISE YOUR SELF-ESTEEM
New York: Bantam, 1987

Brant & Tisza

The Sexually Misused Child
American J. Orthopsychiatry, 47(1)(1977)

Brongersma, Edward

Loving Boys
Amsterdam/New York: Global Academic Publishers, 1987

Bullough & Bullough (Eds.)

Human Sexuality
An Encyclopedia
New York: Garland Publishing, 1994

Sin, Sickness and Sanity
A History of Sexual Attitudes
New York: New American Library, 1977

Buxton, Richard

The Complete World of Greek Mythology
London: Thames & Hudson, 2007

Cain, Chelsea & Moon Unit Zappa

Wild Child
New York: Seal Press (Feminist Publishing), 1999

Calderone & Ramey

Talking With Your Child About Sex
New York: Random House, 1982

Campbell, Herbert James

The Pleasure Areas
London: Eyre Methuen Ltd., 1973

Campbell, Jacqueline C.

Assessing Dangerousness
Violence by Sexual Offenders, Batterers and Child Abusers
New York: Sage Publications, 2004

Campbell, Joseph

The Hero With A Thousand Faces
Princeton: Princeton University Press, 1973
(Bollingen Series XVII)
London: Orion Books, 1999

Occidental Mythology
Princeton: Princeton University Press, 1973
(Bollingen Series XVII)
New York: Penguin Arkana, 1991

The Masks of God
Oriental Mythology
New York: Penguin Arkana, 1992

Originally published 1962

THE POWER OF MYTH
With Bill Moyers
ed. by Sue Flowers
New York: Anchor Books, 1988

CAPACCHIONE, LUCIA

THE POWER OF YOUR OTHER HAND
North Hollywood, CA: Newcastle Publishing, 1988

CAPRA, BERNT AMADEUS

MINDWALK
A Film for Passionate Thinkers
Based Upon Fritjof Capra's The Turning Point
New York: Triton Pictures, 1990

CAPRA, FRITJOF

THE TURNING POINT
Science, Society And The Rising Culture
New York: Simon & Schuster, 1987
Original Author Copyright, 1982

THE TAO OF PHYSICS
An Exploration of the Parallels Between Modern
Physics and Eastern Mysticism
New York: Shambhala Publications, 2000
(New Edition) Originally published in 1975

NATURAL ORDER

The Web of Life
A New Scientific Understanding of Living Systems
New York: Doubleday, 1997
Author Copyright 1996

The Hidden Connections
New York: Doubleday, 2002

Steering Business Toward Sustainability
New York: United Nations University Press, 1995

Uncommon Wisdom
Conversations with Remarkable People
New York: Bantam, 1989

The Science of Leonardo
Inside the Mind of the Great Genius of the Renaissance
New York: Anchor Books, 2008
New York: Bantam Doubleday, 2007 (First Publishing)

Castaneda, Carlos

The Teachings of Don Juan
A Yaqui Way of Knowledge
Washington: Square Press, 1985

Journey to Ixtlan
Washington: Square Press: 1991

Tales of Power
Washington: Square Press, 1991

The Second Ring of Power
Washington: Square Press, 1991

BIBLIOGRAPHY

CLARKE-STEWARD, S., FRIEDMAN, S. & KOCH, J.

CHILD DEVELOPMENT, A TOPICAL APPROACH
London: John Wiley, 1986

CONSTANTINE, LARRY L.

CHILDREN & SEX
New Findings, New Perspectives
Larry L. Constantine & Floyd M. Martinson (Eds.)
Boston: Little, Brown & Company, 1981

TREASURES OF THE ISLAND
Children in Alternative Lifestyles
Beverly Hills: Sage Publications, 1976

WHERE ARE THE KIDS?
in: Libby & Whitehurst (ed.)
Marriage and Alternatives
Glenview: Scott Foresman, 1977

OPEN FAMILY
A Lifestyle for Kids and other People
26 FAMILY COORDINATOR 113-130 (1977)

COOK, M. & HOWELLS, K. (EDS.)

ADULT SEXUAL INTEREST IN CHILDREN
Academic Press, London, 1980

Covitz, Joel

EMOTIONAL CHILD ABUSE
The Family Curse
Boston: Sigo Press, 1986

Currier, Richard L.

JUVENILE SEXUALITY IN GLOBAL PERSPECTIVE
in : Children & Sex, New Findings, New Perspectives
Larry L. Constantine & Floyd M. Martinson (Eds.)
Boston: Little, Brown & Company, 1981

De Bono, Edward

THE USE OF LATERAL THINKING
New York: Penguin, 1967

THE MECHANISM OF MIND
New York: Penguin, 1969

SUR/PETITION
London: HarperCollins, 1993

TACTICS
London: HarperCollins, 1993
First published in 1985

SERIOUS CREATIVITY
Using the Power of Lateral Thinking to Create New Ideas
London: HarperCollins, 1996

BIBLIOGRAPHY

Delacour, Jean-Baptiste

Glimpses of the Beyond
New York: Bantam Dell, 1975

DeMause, Lloyd

The History of Childhood
New York, 1974

Foundations of Psychohistory
New York: Creative Roots, 1982

Diamond, Stephen A., May, Rollo

Anger, Madness, and the Daimonic
The Psychological Genesis of Violence, Evil and Creativity
New York: State University of New York Press, 1999

DiCarlo, Russell E. (Ed.)

Towards A New World View
Conversations at the Leading Edge
Erie, PA: Epic Publishing, 1996

Dolto, Françoise

La Cause des Enfants
Paris: Laffont, 1985

Psychanalyse et Pédiatrie
Paris: Seuil, 1971

SÉMINAIRE DE PSYCHANALYSE D'ENFANTS, 1
Paris: Seuil, 1982

SÉMINAIRE DE PSYCHANALYSE D'ENFANTS, 2
Paris: Seuil, 1985

SÉMINAIRE DE PSYCHANALYSE D'ENFANTS, 3
Paris: Seuil, 1988

L'ÉVANGILE AU RISQUE DE LA PSYCHANALYSE
Paris: Seuil, 1980

DÜRCKHEIM, KARLFRIED GRAF

HARA: THE VITAL CENTER OF MAN
Rochester: Inner Traditions, 2004

ZEN AND US
New York: Penguin Arkana 1991

THE CALL FOR THE MASTER
New York: Penguin Books, 1993

ABSOLUTE LIVING
The Otherworldly in the World and the Path to Maturity
New York: Penguin Arkana, 1992

THE WAY OF TRANSFORMATION
Daily Life as a Spiritual Exercise
London: Allen & Unwin, 1988

THE JAPANESE CULT OF TRANQUILITY
London: Rider, 1960

BIBLIOGRAPHY

EDMUNDS, FRANCIS

AN INTRODUCTION TO ANTHROPOSOPHY
Rudolf Steiner's Worldview
London: Rudolf Steiner Press, 2005

EDWARDES, A.

THE JEWEL OF THE LOTUS
New York, 1959

EINSTEIN, ALBERT

THE WORLD AS I SEE IT
New York: Citadel Press, 1993

OUT OF MY LATER YEARS
New York: Outlet, 1993

IDEAS AND OPINIONS
New York: Bonanza Books, 1988

ALBERT EINSTEIN NOTEBOOK
London: Dover Publications, 1989

EISLER, RIANE

THE CHALICE AND THE BLADE
Our history, Our future
San Francisco: Harper & Row, 1995

SACRED PLEASURE: SEX, MYTH AND THE POLITICS OF THE BODY
New Paths to Power and Love

San Francisco: Harper & Row, 1996

THE PARTNERSHIP WAY
New Tools for Living and Learning
With David Loye
Brandon, VT: Holistic Education Press, 1998

ELWIN, V.

THE MURIA AND THEIR GHOTUL
Bombay: Oxford University Press, 1947

THE SECRET LIFE OF WATER
New York: Atria Books, 2005

ERICKSON, MILTON H.

MY VOICE WILL GO WITH YOU
The Teaching Tales of Milton H. Erickson
by Sidney Rosen (Ed.)
New York: Norton & Co., 1991

COMPLETE WORKS 1.0, CD-ROM
New York: Milton H. Erickson Foundation, 2001

ERIKSON, ERIK H.

CHILDHOOD AND SOCIETY
New York: Norton, 1993
First published in 1950

BIBLIOGRAPHY

Evans-Wentz, Walter Yeeling

The Fairy Faith in Celtic Countries
London: Frowde, 1911
Republished by Dover Publications
(Minneola, New York), 2002

Farson, Richard

Birthrights
A Bill of Rights for Children
Macmillan, New York, 1974

Feinberg, Joel

Harmless Wrongdoing
The Moral Limits of the Criminal Law, Vol. 4
New York: Oxford University Press, 1990

Fensterhalm, Herbert

Don't Say Yes When You Want to Say No
With Jean Bear
New York: Dell, 1980

Finkelhor, David

Sexually Victimized Children
New York: Free Press, 1981

Finkelstein, Haim N. (Ed.)

The Collected Writings of Salvador Dali
Cambridge: Cambridge University Press, 1998

Fortune, Mary M.

Sexual Violence
New York: Pilgrim Press, 1994

Foster/Freed

A Bill of Rights for Children
6 FAMILY LAW QUARTERLY 343 (1972)

Foucault, Michel

The History of Sexuality, Vol. I : The Will to Knowledge
London: Penguin, 1998
First published in 1976

The History of Sexuality, Vol. II : The Use of Pleasure
London: Penguin, 1998
First published in 1984

The History of Sexuality, Vol. III : The Care of Self
London: Penguin, 1998
First published in 1984

BIBLIOGRAPHY

FREUD, SIGMUND

THREE ESSAYS ON THE THEORY OF SEXUALITY
in: The Standard Edition of the Complete Psychological
Works of Sigmund Freud
London: Hogarth Press, 1953-54
Vol. 7, pp. 130 ff
(first published in 1905)

THE INTERPRETATION OF DREAMS
New York: Avon, Reissue Edition, 1980
and in: The Standard Edition of the Complete Psychological
Works of Sigmund Freud , (24 Volumes) ed. by James Strachey
New York: W. W. Norton & Company, 1976

TOTEM AND TABOO
New York: Routledge, 1999
Originally published in 1913

FREUND, KURT

ASSESSMENT OF PEDOPHILIA
in: Cook, M. and Howells, K. (eds.)
Adult Sexual Interest in Children
Academic Press, London, 1980

FROMM, ERICH

THE ANATOMY OF HUMAN DESTRUCTIVENESS
New York: Owl Book, 1992
Originally published in 1973

ESCAPE FROM FREEDOM
New York: Owl Books, 1994
Originally published in 1941

TO HAVE OR TO BE
New York: Continuum International Publishing, 1996
Originally published in 1976

THE ART OF LOVING
New York: HarperPerennial, 2000
Originally published in 1956

GELDARD, RICHARD

REMEMBERING HERACLITUS
New York: Lindisfarne Books, 2000

GERBER, RICHARD

A PRACTICAL GUIDE TO VIBRATIONAL MEDICINE
Energy Healing and Spiritual Transformation
New York: Harper & Collins, 2001

GELLER, URI

THE MINDPOWER KIT
Includes Book, Audiotape, Quartz Crystal And Meditation Circle
New York: Penguin, 1996

GESELL, IZZY

PLAYING ALONG
37 Group Learning Activities Borrowed from Improvisational Theater
Whole Person Associates, 1997

GHISELIN, BREWSTER (ED.)

THE CREATIVE PROCESS
Reflections on Invention in the Arts and Sciences
Berkeley: University of California Press, 1985
First published in 1952

GIBSON, IAN

THE SHAMEFUL LIFE OF SALVADOR DALI
New York: Norton, 1998

GIL, DAVID G.

SOCIETAL VIOLENCE AND VIOLENCE IN FAMILIES
in: David G. Gil, Child Abuse and Violence
New York: Ams Press, 1928

GIMBUTAS, MARIJA

THE LANGUAGE OF THE GODDESS
London: Thames & Hudson, 2001

Goldenstein, Joyce

Einstein: Physicist and Genius
(Great Minds of Science)
New York: Enslow Publishers, 1995

Goldman, Jonathan & Goldman, Andi

Tantra of Sound
Frequencies of Healing
Charlottesville: Hampton Roads, 2005

Healing Sounds
The Power of Harmonies
Rochester: Healing Arts Press, 2002

Healing Sounds
Principles of Sound Healing
DVD, 90 min.
Sacred Mysteries, 2004

Goldstein, Jeffrey H.

Aggression and Crimes of Violence
New York, 1975

Goleman, Daniel

Emotional Intelligence
New York, Bantam Books, 1995

BIBLIOGRAPHY

Gordon, Rosemary

Pedophilia: Normal and Abnormal
in: Kraemer, The Forbidden Love
London, 1976

Gordon Wasson, R.

The Road to Eleusis
Unveiling the Secret of the Mysteries
With Albert Hofmann, Huston Smith, Carl Ruck and Peter Webster
Berkeley, CA: North Atlantic Books, 2008

Goswami, Amit

The Self-Aware Universe
How Consciousness Creates the Material World
New York: Tarcher/Putnam, 1995

Gottlieb, Adam

Peyote and Other Psychoactive Cacti
Ronin Publishing, 2nd edition, 1997

Grof, Stanislav

Ancient Wisdom and Modern Science
New York: State University of New York Press, 1984

BEYOND THE BRAIN
Birth, Death and Transcendence in Psychotherapy
New York: State University of New York, 1985

LSD: DOORWAY TO THE NUMINOUS
The Groundbreaking Psychedelic Research into Realms of the Human Unconscious
Rochester: Park Street Press, 2009

REALMS OF THE HUMAN UNCONSCIOUS
Observations from LSD Research
New York: E.P. Dutton, 1976

THE COSMIC GAME
Explorations of the Frontiers of Human Consciousness
New York: State University of New York Press, 1998

THE HOLOTROPIC MIND
The Three Levels of Human Consciousness
With Hal Zina Bennett
New York: HarperCollins, 1993

WHEN THE IMPOSSIBLE HAPPENS
Adventures in Non-Ordinary Reality
Louisville, CO: Sounds True, 2005

GROTH, A. NICHOLAS

MEN WHO RAPE
The Psychology of the Offender
New York: Perseus Publishing, 1980

HOUSTON, JEAN

THE POSSIBLE HUMAN
A Course in Enhancing Your Physical, Mental, and Creative Abilities
New York: Jeremy P. Tarcher/Putnam, 1982

HOWELLS, KEVIN

ADULT SEXUAL INTEREST IN CHILDREN
Considerations Relevant to Theories of Aetiology in:
Cook, M. and Howells, K. (eds.): Adult Sexual Interest in Children
Academic Press, London, 1980

HUNT, VALERIE

INFINITE MIND
Science of the Human Vibrations of Consciousness
Malibu, CA: Malibu Publishing, 2000

INNOCENTI DECLARATION

DECLARATION ON THE PROTECTION, PROMOTION AND SUPPORT OF BREASTFEEDING
http://www.innocenti15.net/inno.htm

JACKSON, NIGEL

THE RUNE MYSTERIES
With Silver RavenWolf
St. Paul, Minn.: Llewellyn Publications, 2000

Jackson, Stevi

Childhood and Sexuality
New York: Blackwell, 1982

Jaffe, Hans L.C.

Picasso
New York: Abradale Press, 1996

James, William

Writings 1902-1910
The Varieties of Religious Experience / Pragmatism / A Pluralistic Universe / The Meaning of Truth / Some Problems of Philosophy / Essays
New York: Library of America, 1988

Janov, Arthur

Primal Man
The New Consciousness
New York: Crowell, 1975

Johnson, Paul

A History of the Jews
New York: Harper & Row, 1987

BIBLIOGRAPHY

Johnston & Deisher

Contemporary Communal Child Rearing: A First Analysis
52 PEDIATRICS 319 (1973)

Jones, W.H.S., Litt, D.

Pliny Natural History
Cambridge, Mass.: Harvard University Press, 1980

Jung, Carl Gustav

Archetypes of the Collective Unconscious
in: The Basic Writings of C.G. Jung
New York: The Modern Library, 1959, 358-407

Collected Works
New York, 1959

On the Nature of the Psyche
in: The Basic Writings of C.G. Jung
New York: The Modern Library, 1959, 47-133

Psychological Types
Collected Writings, Vol. 6
Princeton: Princeton University Press, 1971

Psychology and Religion
in: The Basic Writings of C.G. Jung
New York: The Modern Library, 1959, 582-655

Religious and Psychological Problems of Alchemy
in: The Basic Writings of C.G. Jung
New York: The Modern Library, 1959, 537-581

SYMBOL UND LIBIDO
Freiburg: Walter Verlag, 1987

THE BASIC WRITINGS OF C.G. JUNG
New York: The Modern Library, 1959

THE DEVELOPMENT OF PERSONALITY
Collected Writings, Vol. 17
Princeton: Princeton University Press, 1954

THE MEANING AND SIGNIFICANCE OF DREAMS
Boston: Sigo Press, 1991

THE MYTH OF THE DIVINE CHILD
in: Essays on A Science of Mythology
Princeton, N.J.: Princeton University Press Bollingen Series XXII, 1969. (With Karl Kerenyi)

TWO ESSAYS ON ANALYTICAL PSYCHOLOGY
Collected Writings, Vol. 7
Princeton: Princeton University Press, 1972
First published by Routledge & Kegan Paul, Ltd., 1953

KAHN, CHARLES (ED.)

THE ART AND THOUGHT OF HERACLITUS
Cambridge: Cambridge University Press, 2008

KAPLEAU, ROSHI PHILIP

THREE PILLARS OF ZEN
Boston: Beacon Press, 1967

BIBLIOGRAPHY

KARAGULLA, SHAFICA

THE CHAKRAS
Correlations between Medical Science and Clairvoyant Observation (With Dora van Gelder Kunz)
Wheaton: Quest Books, 1989

KLEIN, MELANIE

LOVE, GUILT AND REPARATION, AND OTHER WORKS 1921-1945
New York: Free Press, 1984
(Reissue Edition)

ENVY AND GRATITUDE AND OTHER WORKS 1946-1963
New York: Free Press, 2002
(Reissue Edition)

KRAEMER

THE FORBIDDEN LOVE
London, 1976

KRAFFT-EBING, RICHARD VON

PSYCHOPATHIA SEXUALIS
New York: Bell Publishing, 1965
Originally published in 1886

Krause, Donald G.

The Art of War for Executives
London: Nicholas Brealey Publishing, 1995

Krishnamurti, J.

Freedom From The Known
San Francisco: Harper & Row, 1969

The First and Last Freedom
San Francisco: Harper & Row, 1975

Education and the Significance of Life
London: Victor Gollancz, 1978

Commentaries on Living
First Series
London: Victor Gollancz, 1985

Commentaries on Living
Second Series
London: Victor Gollancz, 1986

Krishnamurti's Journal
London: Victor Gollancz, 1987

Krishnamurti's Notebook
London: Victor Gollancz, 1986

Beyond Violence
London: Victor Gollancz, 1985

Beginnings of Learning
New York: Penguin, 1986

BIBLIOGRAPHY

THE PENGUIN KRISHNAMURTI READER
New York: Penguin, 1987

ON GOD
San Francisco: Harper & Row, 1992

ON FEAR
San Francisco: Harper & Row, 1995

THE ESSENTIAL KRISHNAMURTI
San Francisco: Harper & Row, 1996

THE ENDING OF TIME
With Dr. David Bohm
San Francisco: Harper & Row, 1985

LAING, RONALD DAVID

DIVIDED SELF
New York: Viking Press, 1991

R.D. LAING AND THE PATHS OF ANTI-PSYCHIATRY
ed., by Z. Kotowicz
London: Routledge, 1997

THE POLITICS OF EXPERIENCE
New York: Pantheon, 1983

LAKHOVSKY, GEORGES

SECRET OF LIFE
New York: Kessinger Publishing, 2003

LASZLO, ERVIN

SCIENCE AND THE AKASHIC FIELD
An Integral Theory of Everything
Rochester: Inner Traditions, 2004

QUANTUM SHIFT TO THE GLOBAL BRAIN
How the New Scientific Reality Can Change Us and Our World
Rochester: Inner Traditions, 2008

SCIENCE AND THE REENCHANTMENT OF THE COSMOS
The Rise of the Integral Vision of Reality
Rochester: Inner Traditions, 2006

THE AKASHIC EXPERIENCE
Science and the Cosmic Memory Field
Rochester: Inner Traditions, 2009

THE CHAOS POINT
The World at the Crossroads
Newburyport, MA: Hampton Roads Publishing, 2006

LAUD, ANNE & GILSTROP, MAY

VIOLENCE IN THE FAMILY
A Selected Bibliography on Child Abuse, Sexual Abuse of Children & Domestic Violence, June 1985, University of Georgia Libraries, Bibliographical Series, No. 32

LEADBEATER, CHARLES WEBSTER

ASTRAL PLANE
Its Scenery, Inhabitants and Phenomena
Kessinger Publishing Reprint Edition, 1997

BIBLIOGRAPHY

DREAMS
What they Are and How they are Caused
London: Theosophical Publishing Society, 1903
Kessinger Publishing Reprint Edition, 1998

THE INNER LIFE
Chicago: The Rajput Press, 1911
Kessinger Publishing

LEARY, TIMOTHY

OUR BRAIN IS GOD
Berkeley, CA: Ronin Publishing, 2001
Author Copyright 1988

LEBOYER, FREDERICK

BIRTH WITHOUT VIOLENCE
New York, 1975

INNER BEAUTY, INNER LIGHT
New York: Newmarket Press, 1997

LOVING HANDS
The Traditional Art of Baby Massage
New York: Newmarket Press, 1977

THE ART OF BREATHING
New York: Newmarket Press, 1991

LEGGETT, TREVOR P.

A First Zen Reader
Rutland: C.E. Tuttle, 1980
Originally published in 1972

LEONARD, GEORGE, MURPHY, MICHAEL

The Live We Are Given
A Long Term Program for Realizing the
Potential of Body, Mind, Heart and Soul
New York: Jeremy P. Tarcher/Putnam, 1984

LICHT, HANS

Sexual Life in Ancient Greece
New York: AMS Press, 1995

LIEDLOFF, JEAN

Continuum Concept
In Search of Happiness Lost
New York: Perseus Books, 1986
First published in 1977

LIPTON, BRUCE

The Biology of Belief
Unleashing the Power of Consciousness, Matter and Miracles
Santa Rosa, CA: Mountain of Love/Elite Books, 2005

BIBLIOGRAPHY

Locke, John

Some Thoughts Concerning Education
London, 1690
Reprinted in: The Works of John Locke, 1823
Vol. IX., pp. 6-205

Long, Max Freedom

The Secret Science at Work
The Huna Method as a Way of Life
Marina del Rey: De Vorss Publications, 1995
Originally published in 1953

Growing Into Light
A Personal Guide to Practicing the Huna Method,
Marina del Rey: De Vorss Publications, 1955

Lowen, Alexander

Bioenergetics
New York: Coward, McGoegham 1975

Depression and the Body
The Biological Basis of Faith and Reality
New York: Penguin, 1992

Fear of Life
New York: Bioenergetic Press, 2003

Honoring the Body
The Autobiography of Alexander Lowen
New York: Bioenergetic Press, 2004

Joy
The Surrender to the Body and to Life
New York: Penguin, 1995

Love and Orgasm
New York: Macmillan, 1965

Love, Sex and Your Heart
New York: Bioenergetics Press, 2004

Narcissism: Denial of the True Self
New York: Macmillan, Collier Books, 1983

Pleasure: A Creative Approach to Life
New York: Bioenergetics Press, 2004
First published in 1970

The Language of the Body
Physical Dynamics of Character Structure
New York: Bioenergetics Press, 2006

Malinowski, Bronislaw

Crime und Custom in Savage Society
London: Kegan, 1926

Sex and Repression in Savage Society
London: Kegan, 1927

The Sexual Life of Savages in North West Melanesia
New York: Halycon House, 1929

MANN, EDWARD W.

ORGONE, REICH & EROS
Wilhelm Reich's Theory of Life Energy
New York: Simon & Schuster (Touchstone), 1973

MARTINSON, FLOYD M.

SEXUAL KNOWLEDGE
Values and Behavior Patterns
St. Peter: Minn.: Gustavus Adolphus College, 1966

INFANT AND CHILD SEXUALITY
St. Peter: Minn.: Gustavus Adolphus College, 1973

THE QUALITY OF ADOLESCENT EXPERIENCES
St. Peter: Minn.: Gustavus Adolphus College, 1974

THE CHILD AND THE FAMILY
Calgary, Alberta: The University of Calgary, 1980

THE SEX EDUCATION OF YOUNG CHILDREN
in: Lorna Brown (Ed.), Sex Education in the Eighties
New York, London: Plenum Press, 1981, pp. 51 ff.

THE SEXUAL LIFE OF CHILDREN
New York: Bergin & Garvey, 1994

CHILDREN AND SEX, PART II: CHILDHOOD SEXUALITY
in: Bullough & Bullough, Human Sexuality (1994)
Pp. 111-116

Masters, R.E.L.

Forbidden Sexual Behavior and Morality
New York, 1962

McCarey, William A.

In Search of Healing
Whole-Body Healing Through the Mind-Body-Spirit Connection
New York: Berkley Publishing, 1996

McLeod, Kembrew

Freedom of Expression
Resistance and Repression in the Age of Intellectual Property
Minneapolis, MN: University of Minnesota Press, 2007

McTaggart, Lynne

The Field
The Quest for the Secret Force of the Universe
New York: Harper & Collins, 2002

Mead, Margaret

Sex and Temperament in Three Primitive Societies
New York, 1935

BIBLIOGRAPHY

MEADOWS, DONELLA H.

THINKING IN SYSTEMS
A Primer
White River, VT: Chelsea Green Publishing, 2008

MEHTA, ROHIT

J. KRISHNAMURTI AND THE NAMELESS EXPERIENCE
A Comprehensive Discussion of J. Krishnamurti's Approach to Life
Delhi: Motilal Banarsidass Publishers, 2002

MERLEAU-PONTY, MAURICE

PHENOMENOLOGY OF PERCEPTION
London: Routledge, 1995
Originally published 1945

METZNER, RALPH (ED.)

AYAHUASCA, HUMAN CONSCIOUSNESS AND THE SPIRITS OF NATURE
ed. by Ralph Metzner, Ph.D
New York: Thunder's Mouth Press, 1999

THE PSYCHEDELIC EXPERIENCE
A Manual Based on the Tibetan Book of the Dead
With Timothy Leary and Richard Alpert
New York: Citadel, 1995

Miller, Alice

For Your Own Good
Hidden Cruelty in Child-Rearing and the Roots of Violence
New York: Farrar, Straus & Giroux, 1983

Pictures of a Childhood
New York: Farrar, Straus & Giroux, 1986

The Drama of the Gifted Child
In Search for the True Self
translated by Ruth Ward
New York: Basic Books, 1996

Thou Shalt Not Be Aware
Society's Betrayal of the Child
New York: Noonday, 1998

The Political Consequences of Child Abuse
in: The Journal of Psychohistory 26, 2 (Fall 1998)

Moll, Albert

The Sexual Life of the Child
New York: Macmillan, 1912
First published in German as
Das Sexualleben des Kindes, 1909

Monroe, Robert

Ultimate Journey
New York: Broadway Books, 1994

MONTAGU, ASHLEY

TOUCHING
The Human Significance of the Skin
New York: Harper & Row, 1978

MONTESSORI, MARIA

THE ABSORBENT MIND
Reprint Edition
New York: Buccaneer Books, 1995
First published in 1973

MOORE, THOMAS

CARE OF THE SOUL
A Guide for Cultivating Depth and Sacredness in Everyday Life
New York: Harper & Collins, 1994

MOSER, CHARLES ALLEN

DSM-IV-TR AND THE PARAPHILIAS: AN ARGUMENT FOR REMOVAL
With Peggy J. Kleinplatz
Journal of Psychology and Human Sexuality 17 (3/4), 91-109
(2005)

MURDOCK, G.

SOCIAL STRUCTURE
New York: Macmillan, 1960

Murphy, Joseph

The Power of Your Subconscious Mind
West Nyack, N.Y.: Parker, 1981, N.Y.: Bantam, 1982
Originally published in 1962

The Miracle of Mind Dynamics
New York: Prentice Hall, 1964

Miracle Power for Infinite Riches
West Nyack, N.Y.: Parker, 1972

The Amazing Laws of Cosmic Mind Power
West Nyack, N.Y.: Parker, 1973

Secrets of the I Ching
West Nyack, N.Y.: Parker, 1970

Think Yourself Rich
Use the Power of Your Subconscious Mind to Find True Wealth
Revised by Ian D. McMahan, Ph.D.
Paramus, NJ: Reward Books, 2001

Murphy, Michael

The Future of the Body
Explorations into the Further Evolution of Human Nature
New York: Jeremy P. Tarcher/Putnam, 1992

Myers, Tony Pearce

The Soul of Creativity
Insights into the Creative Process
Novato, CA: New World Library, 1999

Myss, Caroline

The Creation of Health
The Emotional, Psychological, and Spiritual Responses that Promote Health and Healing
New York: Three Rivers Press, 1998

Naparstek, Belleruth

Your Sixth Sense
Unlocking the Power of Your Intuition
London: HarperCollins, 1998

Staying Well With Guided Imagery
New York: Warner Books, 1995

Narby, Jeremy

The Cosmic Serpent
DNA and the Origins of Knowledge
New York: J. P. Tarcher, 1999

Nau, Erika

Self-Awareness Through Huna
Virginia Beach: Donning, 1981

Neill, Alexander Sutherland

Neill! Neill! Orange-Peel!
New York: Hart Publishing Co., 1972

SUMMERHILL
A Radical Approach to Child Rearing
New York: Hart Publishing, Reprint 1984
Originally published 1960

SUMMERHILL SCHOOL
A New View of Childhood
New York: St. Martin's Press
Reprint 1995

NEUMANN, ERICH

THE GREAT MOTHER
Princeton: Princeton University Press, 1955
(Bollingen Series)

NEWTON, MICHAEL

LIFE BETWEEN LIVES
Hypnotherapy for Spiritual Regression
Woodbury, Minn.: Llewellyn Publications, 2006

NICHOLS, SALLIE

JUNG AND TAROT: AN ARCHETYPAL JOURNEY
New York: Red Wheel/Weiser, 1986

NIN, ANAÏS

THE DIARY OF ANAÏS NIN (7 VOLUMES)
New York, 1966

BIBLIOGRAPHY

Volume 1 (1931-1934)
New York: Harvest Books, 1969

Volume 2 (1934-1939)
New York: Harvest Books, 1970

Odent, Michel

Birth Reborn
What Childbirth Should Be
London: Souvenir Press, 1994

The Scientification of Love
London: Free Association Books, 1999

Primal Health
Understanding the Critical Period Between Conception
and the First Birthday
London: Clairview Books, 2002
First Published in 1986 with Century Hutchinson in London

The Functions of the Orgasms
The Highway to Transcendence
London: Pinter & Martin, 2009

Ollendorf-Reich, Ilse

Wilhelm Reich, A Personal Biography
New York, St. Martins Press, 1969

Wilhelm Reich
Vorwort von A.S. Neill
München, Kindler, 1975

Pearce Myers, Tony (Editor)

The Soul of Creativity
Insights into the Creative Process
Novato: New World Library, 1999

Pert, Candace B.

Molecules of Emotion
The Science Behind Mind-Body Medicine
New York: Scribner, 2003

Petrash, Jack

Understanding Waldorf Education
Teaching from the Inside Out
London: Floris Books, 2003

Plummer, Kenneth

Pedophilia
Constructing a Sociological Baseline
in: in: Cook, M. and Howells, K. (Eds.):
Adult Sexual Interest in Children
Academic Press, London, 1980, pp. 220 ff.

Porteous, Hedy S.

Sex and Identity
Your Child's Sexuality
Indianapolis: Bobbs-Merrill, 1972

BIBLIOGRAPHY

PRESCOTT, JAMES W.

AFFECTIONAL BONDING FOR THE PREVENTION OF VIOLENT BEHAVIORS
Neurobiological, Psychological and Religious/Spiritual Determinants, in: Hertzberg, L.J., Ostrum, G.F. and Field, J.R., (Eds.)

VIOLENT BEHAVIOR
Vol. 1, Assessment & Intervention, Chapter Six
New York: PMA Publishing, 1990

ALIENATION OF AFFECTION
Psychology Today, December 1979

BODY PLEASURE AND THE ORIGINS OF VIOLENCE
Bulletin of the Atomic Scientists, 10-20 (1975)

DEPRIVATION OF PHYSICAL AFFECTION AS A PRIMARY PROCESS IN THE DEVELOPMENT OF PHYSICAL VIOLENCE A COMPARATIVE AND CROSS-CULTURAL PERSPECTIVE, IN: DAVID G. GIL, ED., CHILD ABUSE AND VIOLENCE
New York: Ams Press, 1979

EARLY SOMATOSENSORY DEPRIVATION AS AN ONTOGENETIC PROCESS IN THE ABNORMAL DEVELOPMENT OF THE BRAIN AND BEHAVIOR,
in: Medical Primatology, ed. by I.E. Goldsmith and J. Moor-Jankowski,
New York: S. Karger, 1971

GENITAL MUTILATION OF CHILDREN: FAILURE OF HUMANITY AND HUMANISM
Unprinted Essay (2005)
http://www.violence.de/prescott/letters/CIRC_CONGRESS_MONTAGUE_9.30.05.html

GENITAL PAIN VS. GENITAL PLEASURE
Why the One and not the Other

The Truth Seeker, July/August 1989, pp. 14-21
http://www.violence.de/prescott/truthseeker/genpl.html

HOW CULTURE SHAPES THE DEVELOPING BRAIN AND THE FUTURE OF HUMANITY
A Brief Summary of the research which links brain abnormalities and violence to an absence of nurturing and bonding very early in childhood, in: Touch the Future: Optimum Learning Relationships

FOR CHILDREN & ADULTS
Spring 2002 (Ed. by Michael Mendizza)
Nevada City, CA, 2002

INVITED COMMENTARY: CENTRAL NERVOUS SYSTEM FUNCTIONING IN ALTERED SENSORY ENVIRONMENTS
in: M.H. Appley and R. Trumbull (Eds.), Psychological Stress,
New York: Appleton-Century Crofts, 1967

OUR TWO CULTURAL BRAINS: NEUROINTEGRATIVE AND NEURODISSOCIATIVE
http://www.violence.de/prescott/letters/Our_Two_Cultural_Brains.pdf

PHYLOGENETIC AND ONTOGENETIC ASPECTS OF HUMAN AFFECTIONAL DEVELOPMENT,
in: Progress in Sexology, Proceedings of the 1976 International, Congress of Sexology, ed. by R. Gemme & C.C. Wheeler, New York: Plenum Press, 1977

PREVENTION OR THERAPY AND THE POLITICS OF TRUST INSPIRING A NEW HUMAN AGENDA
in: Psychotherapy and Politics International
Volume 3(3), pp. 194-211
London: John Wiley, 2005

SEX AND THE BRAIN
Midcontinent & Eastern Regions, June 13-16, 2002

BIBLIOGRAPHY

Big Rapids, MI: Society for Cross-Cultural Research,
32nd Annual Meeting, 2005
http://www.violence.de/archive.shtml

SIXTEEN PRINCIPLES FOR PERSONAL, FAMILY AND GLOBAL PEACE
The Truth Seeker, March/April 1989
http://www.violence.de/prescott/letters/Sixteen_Principles.pdf

SOMATOSENSORY AFFECTIONAL DEPRIVATION (SAD) THEORY OF DRUG AND ALCOHOL USE
in: Theories on Drug Abuse: Selected Contemporary Perspectives, ed. by Dan J. Lettieri, Mollie Sayers and Helen Wallenstien Pearson, NIDA Research Monograph 30, March 1980, Rockville, MD: National Institute on Drug Abuse, Department of Health and Human Services, 1980

THE ORIGINS OF HUMAN LOVE AND VIOLENCE
Pre- and Perinatal Psychology Journal, Volume 10, Number 3: Spring 1996, pp. 143-188The Origins of Love and Violence

SENSORY DEPRIVATION AND THE DEVELOPING BRAIN
Research and Prevention (DVD)
http://ttfuture.org/store/origins_orders
http://violence.de
http://ttfuture.org/violence
http://montagunocircpetition.org

PRITCHARD, COLIN

THE CHILD ABUSERS
New York: Open University Press, 2004

Raknes, Ola

Wilhelm Reich and Orgonomy
Oslo: Universitetsforlaget, 1970

Randall, Neville

Life After Death
London: Robert Hale, 1999

Rank, Otto

Art and Artist
With Charles Francis Atkinson and Anaïs Nin
New York: W.W. Norton, 1989
Originally published in 1932

The Significance of Psychoanalysis for the Mental Sciences
New York: BiblioBazaar, 2009
First published in 1913

Redfield, James

The Tenth Insight
Holding the Vision
New York: Warner Books, 1996

The Celestine Prophecy
New York: Warner Books, 1995

BIBLIOGRAPHY

Reich, Wilhelm

A REVIEW OF THE THEORIES, DATING FROM THE 17TH CENTURY, ON THE ORIGIN OF ORGANIC LIFE
by Arthur Hahn, Literature Assistant at the Institut für Sexualökonomische Lebensforschung, Biologisches Laboratorium, Oslo, 1938, ©1979 Mary Boyd Higgins as Director of the Wilhelm Reich Infant Trust, XEROX Copy from the Wilhelm Reich Museum

CHILDREN OF THE FUTURE
On the Prevention of Sexual Pathology
New York: Farrar, Straus & Giroux, 1984
First published in 1950

CORE (COSMIC ORGONE ENGINEERING)
Part I, Space Ships, DOR and DROUGHT
©1984, Orgone Institute Press
XEROX Copy from the Wilhelm Reich Museum
Köln: Kiepenheuer & Witsch, 1987

EARLY WRITINGS 1
New York: Farrar, Straus & Giroux, 1975

ETHER, GOD & DEVIL & COSMIC SUPERIMPOSITION
New York: Farrar, Straus & Giroux, 1972
Originally published in 1949

GENITALITY IN THE THEORY AND THERAPY OF NEUROSIS
©1980 by Mary Boyd Higgins as Director of the Wilhelm Reich Infant Trust

PEOPLE IN TROUBLE
©1974 by Mary Boyd Higgins as Director of the Wilhelm Reich Infant Trust

NATURAL ORDER

RECORD OF A FRIENDSHIP
The Correspondence of Wilhelm Reich and A. S. Neill
New York, Farrar, Straus & Giroux, 1981

SELECTED WRITINGS
An Introduction to Orgonomy
New York: Farrar, Straus & Giroux, 1973

THE BIOELECTRICAL INVESTIGATION OF SEXUALITY AND ANXIETY
New York: Farrar, Straus & Giroux, 1983
Originally published in 1935

THE BION EXPERIMENTS
reprinted in Selected Writings
New York: Farrar, Straus & Giroux, 1973

THE CANCER BIOPATHY (THE ORGONE, VOL. 2)
New York: Farrar, Straus & Giroux, 1973

THE FUNCTION OF THE ORGASM (THE ORGONE, VOL. 1)
Orgone Institute Press, New York, 1942

THE INVASION OF COMPULSORY SEX MORALITY
New York: Farrar, Straus & Giroux, 1971
Originally published in 1932

THE LEUKEMIA PROBLEM: APPROACH
©1951, Orgone Institute Press
Copyright Renewed 1979
XEROX Copy from the Wilhelm Reich Museum

THE MASS PSYCHOLOGY OF FASCISM
New York: Farrar, Straus & Giroux, 1970
Originally published in 1933

THE ORGONE ENERGY ACCUMULATOR
Its Scientific and Medical Use

BIBLIOGRAPHY

©1951, 1979, Orgone Institute Press
XEROX Copy from the Wilhelm Reich Museum

THE SCHIZOPHRENIC SPLIT
©1945, 1949, 1972 by Mary Boyd Higgins as Director of the Wilhelm Reich Infant Trust
XEROX Copy from the Wilhelm Reich Museum

THE SEXUAL REVOLUTION
©1945, 1962 by Mary Boyd Higgins as Director of the Wilhelm Reich Infant Trust

RISO, DON RICHARD & HUDSON, RUSS

THE WISDOM OF THE ENNEAGRAM
The Complete Guide to Psychological and Spiritual Growth For The Nine Personality Types
New York: Bantam Books, 1999

ROBBINS, ANTHONY

AWAKEN THE GIANT WITHIN
New York: Simon & Schuster, 1991

UNLIMITED POWER
The New Science of Personal Achievement
New York: Free Press, 1997

ROBERTS, JANE

THE NATURE OF PERSONAL REALITY
New York: Amber-Allen Publishing, 1994
First published in 1974

THE NATURE OF THE PSYCHE
Its Human Expression
New York, Amber-Allen Publishing, 1996
First published in 1979

ROSEN, SYDNEY (ED.)

MY VOICE WILL GO WITH YOU
The Teaching Tales of Milton H. Erickson
New York: Norton & Co., 1991

ROTHSCHILD & WOLF

CHILDREN OF THE COUNTERCULTURE
New York: Garden City, 1976

SANDFORT, THEO

THE SEXUAL ASPECT OF PEDOPHILE RELATIONS
The Experience of Twenty-five Boys
Amsterdam: Pan/Spartacus, 1982

SCHLIPP, PAUL A. (ED.)

ALBERT EINSTEIN
Philosopher-Scientist
New York: Open Court Publishing, 1988

BIBLIOGRAPHY

SCHWARTZ, ANDREW E.

GUIDED IMAGERY FOR GROUPS
Fifty Visualizations That Promote Relaxation, Problem-Solving, Creativity, and Well-Being
Whole Person Associates, 1995

SHARAF, MYRON

FURY ON EARTH
A Biography of Wilhelm Reich
London: André Deutsch, 1983

SHELDRAKE, RUPERT

A NEW SCIENCE OF LIFE
The Hypothesis of Morphic Resonance
Rochester: Park Street Press, 1995

SHER, BARBARA & GOTTLIEB, ANNIE

WISHCRAFT
How to Get What You Really Want
2nd edition, New York: Ballantine Books, 2003

SHONE, RONALD

CREATIVE VISUALIZATION
Using Imagery and Imagination for Self-Transformation
New York: Destiny Books, 1998

SIMONTON, O. CARL ET AL.

GETTING WELL AGAIN
Los Angeles: Tarcher, 1978

SINGER, JUNE

ANDROGYNY
New York: Doubleday Dell, 1976

SMITH, C. MICHAEL

JUNG AND SHAMANISM IN DIALOGUE
London: Trafford Publishing, 2007

SPOCK, BENJAMIN

DR. SPOCK'S BABY AND CHILD CARE
8th Edition
New York: Pocket Books, 2004

STEIN, ROBERT M.

REDEEMING THE INNER CHILD IN MARRIAGE AND THERAPY
in: Reclaiming the Inner Child
ed. by Jeremiah Abrams
New York: Tarcher/Putnam, 1990, 261 ff.

BIBLIOGRAPHY

Steiner, Rudolf

THEOSOPHY
An Introduction to the Spiritual Processes in Human Life and in the Cosmos
New York: Anthroposophic Press, 1994

Stekel, Wilhelm

AUTO-EROTICISM
A Psychiatric Study of Onanism and Neurosis
Republished, London: Paul Kegan, 2004

PATTERNS OF PSYCHOSEXUAL INFANTILISM
New York, 1959 (reprint edition)

SADISM AND MASOCHISM
New York: W.W. Norton & Co., 1953

SEX AND DREAMS
The Language of Dreams
Republished
New York: University Press of the Pacific, 2003

Stiene, Bronwen & Frans

THE REIKI SOURCEBOOK
New York: O Books, 2003

THE JAPANESE ART OF REIKI
A Practical Guide to Self-Healing
New York: O Books, 2005

Stone, Hal & Stone, Sidra

Embracing Our Selves
The Voice Dialogue Manual
San Rafael, CA: New World Library, 1989

Strassman, Rick

DMT: The Spirit Molecule
A doctor's revolutionary research into the biology of near-death and mystical experiences
Rochester: Park Street Press, 2001

Symonds, John Addington

A Problem in Greek Ethics
New York: M.S.G. House, 1971

Szasz, Thomas

The Myth of Mental Illness
New York: Harper & Row, 1984

Talbot, Michael

The Holographic Universe
New York: HarperCollins, 1992

BIBLIOGRAPHY

TARNAS, RICHARD

COSMOS AND PSYCHE
Intimations of a New World View
New York: Plume, 2007

THE PASSION OF THE WESTERN MIND
Understanding the Ideas that have Shaped Our World View
New York: Ballantine Books, 1993

TART, CHARLES T.

ALTERED STATES OF CONSCIOUSNESS
A Book of Readings
Hoboken, N.J.: Wiley & Sons, 1969

TEXTOR, R. B.

A CROSS-CULTURAL SUMMARY
New Haven, Human Relations Area Files (HRAF)
Press, 1967

THE ADVENT OF GREAT AWAKENING

A COURSE IN MIRACLES
Text Workbook and Manual for Teachers
New York: New Christian Church of Full Endeavor, 2007

TILLER, WILLIAM A.

CONSCIOUS ACTS OF CREATION
The Emergence of a New Physics
Associated Producers, 2004 (DVD)

PSYCHOENERGETIC SCIENCE
New York: Pavior, 2007

TOFFLER, ALVIN

POWERSHIFT
Knowledge, Wealth, and Violence at the Edge of the 21st Century
New York: Bantam, 1991

REVOLUTIONARY WEALTH
How it will be created and how it will change our lives
New York: Broadway Business, 2007

THE THIRD WAVE
New York: Bantam, 1984

TOLLE, ECKHART

THE POWER OF NOW
A Guide to Spiritual Enlightenment
Novato, CA: New World Library, 2004

A NEW EARTH: AWAKENING TO YOUR LIFE'S PURPOSE
New York: Michael Joseph (Penguin), 2005

BIBLIOGRAPHY

Van GElder, Dora

The Real World of Fairies
A First-Person Account
2nd Edition
Wheaton: Quest Books, 1999

Villoldo, Alberto

Healing States
A Journey Into the World of Spiritual Healing and Shamanism
With Stanley Krippner
New York: Simon & Schuster (Fireside), 1987

Dance of the Four Winds: Secrets of the Inca Medicine Wheel
With Eric Jendresen
Rochester: Destiny Books, 1995

Shaman, Healer, Sage
How to Heal Yourself and Others with the Energy Medicine
of the Americas
New York: Harmony, 2000

Healing the Luminous Body
The Way of the Shaman with Dr. Alberto Villoldo
DVD, Sacred Mysteries Productions, 2004

Mending The Past And Healing The Future with Soul Retrieval
New York: Hay House, 2005

Ward, Elizabeth

Father-Daughter Rape
New York: Grove Press, 1985

WHITFIELD, CHARLES L.

HEALING THE CHILD WITHIN
Deerfield Beach, Fl: Health Communications, 1987

WHITING, BEATRICE B.

CHILDREN OF SIX CULTURES
A Psycho-Cultural Analysis
Cambridge: Harvard University Press, 1975

WILBER, KEN

SEX, ECOLOGY, SPIRITUALITY
The Spirit of Evolution
Boston: Shambhala, 2000

QUANTUM QUESTIONS
Mystical Writings of The World's Greatest Physicists
Boston: Shambhala, 2001

WILLIAMS, STREPHON KAPLAN

DREAMS AND SPIRITUAL GROWTH
With Patricia H. Berne and Louis M. Savary
New York: Paulist Press, 1984

DREAM CARDS
Understand Your Dreams and Enrich Your Life
New York: Simon & Schuster (Fireside), 1991

BIBLIOGRAPHY

WOLF, FRED ALAN

TAKING THE QUANTUM LEAP
The New Physics for Nonscientists
New York: Harper & Row, 1989

PARALLEL UNIVERSES
New York: Simon & Schuster, 1990

THE DREAMING UNIVERSE
A Mind-Expanding Journey into the Realm Where Psyche and Physics Meet
New York: Touchstone, 1995

THE EAGLE'S QUEST
A Physicist Finds the Scientific Truth At the Heart of the Shamanic World
New York: Touchstone, 1997

YATES, ALAYNE

SEX WITHOUT SHAME: ENCOURAGING THE CHILD'S HEALTHY SEXUAL DEVELOPMENT
New York, 1978
Republished Internet Edition

ZUKAV, GARY

THE DANCING WU LI MASTERS
An Overview of the New Physics
New York: HarperOne, 2001

Personal Notes

www.ingramcontent.com/pod-product-compliance
Lightning Source LLC
Chambersburg PA
CBHW020624220526
45464CB00001B/9